T0076303

VULTURES OF THE WORLD

VULTURES OF THE WORLD

ESSENTIAL ECOLOGY AND CONSERVATION

Keith L. Bildstein

Comstock Publishing Associates
an imprint of
Cornell University Press
Ithaca and London

First published 2022 by Cornell University Press

Printed in the United States of America

Library of Congress Cataloging-in-Publication Data

Names: Bildstein, Keith L., author.
Title: Vultures of the world : essential ecology and conservation / Keith L. Bildstein.
Description: Ithaca [New York] : Comstock Publishing Associates, an imprint of Cornell University Press, 2022. | Includes bibliographical references and index.
Identifiers: LCCN 2021041876 (print) | LCCN 2021041877 (ebook) | ISBN 9781501761614 (hardcover) | ISBN 9781501765025 (pdf) | ISBN 9781501765032 (epub)
Subjects: LCSH: Vultures—Ecology. | Condors—Ecology. | Vultures—Conservation. | Condors—Conservation.
Classification: LCC QL696.F32 B55 2022 (print) | LCC QL696.F32 (ebook) | DDC 598.9/2—dc23
LC record available at https://lccn.loc.gov/2021041876
LC ebook record available at https://lccn.loc.gov/2021041877

To Lindy, Corrine, Andre, Campbell,
Darcy, Munir, Peter, Sergio, and Todd
whose love for vultures
proved to be contagious

CONTENTS

PREFACE

VULTURES OF THE WORLD owes its origins to a chance encounter at Hawk Mountain Sanctuary in the Appalachian Mountains of eastern Pennsylvania, USA, in the autumn of 2002. Two Princeton University scientists, David Wilcove and Martin Wikelski, were visiting the Sanctuary to take in the autumn raptor migration when I encountered them at the Sanctuary's South Lookout while interpreting the flight for visitors. After discussing several of the "big questions" in raptor movement ecology, the three of us agreed that an in-depth study of the migrations of Turkey Vultures was long overdue. Doing so made sense for several reasons. First, Turkey Vultures were "partial migrants," a migratory species in which some but not all individuals migrated, and although partial migrants were far more common than "complete migrants," they remained less studied and the geographic complexities of their movements were less well understood than for the complete migrants. Second, Turkey Vulture populations were doing well at a time when most vultures were not, which offered the chance to study the ecology of a conservation outlier. Third, Turkey Vultures were both common and widespread in the Americas and could be observed in the backyards and, perhaps more important, the schoolyards of most children in the New World, making them accessible wildlife to both parents and educators throughout their range. Indeed, the species was common enough in New Jersey, USA, when I was growing up that I cannot recall when I saw my first Turkey Vulture, as they were as common in the sky as clouds.

As we concluded our conversation, Martin Wikelski promised to introduce me to one of his star students, Jamie Mendel, who had just committed to graduate school at Cornell University and was looking for a suitable study project. Less than a year later, Jamie and I were trapping Turkey Vultures; fitting them with satellite-tracking devices, heart-rate monitors, and core-body-temperature dataloggers; and following their movements and tracking metabolic costs. Since then my studies have expanded to include the movement behavior and ecology of migratory European vultures overwintering in Africa, as well the ecology of non-migratory Hooded Vultures in West, East, and southern Africa.

Like other raptor biologists, I have been fortunate in having patient mentors and, subsequently, colleagues who have been willing to share and retell their successful and unsuccessful experiences with me, not only while in the field but also later in the day over food and drink. I have carried on this tradition with my own students and colleagues, and much of what I have learned about vulture and condor biology results from these associations. One of my more memorable mentors was Frances Hamerstrom, a tough-as-nails, old-school role model who shepherded me, sometimes begrudgingly, through the whys and wherefores of what she would have called an introduction to conducting "proper animal ecology." Hamerstrom had a series of rules for doing this successfully, three of which stand out. First and foremost was to understand that you were largely ignorant of your subject, and that there was always more to be learned. Second was that thoroughly understanding and appreciating the fundamental principles of ecology were essential. Third was that among the most important of these principles were the processes involved in population limitation. This book reflects those rules as it presents arguments for understanding the essential nature of vultures, the world's scavenging birds of prey.

Vultures of the World begins with an introduction to the evolutionary processes that have led to the global diversity of scavenging birds of prey, along with a framework for discussing the ecological aspects necessary for sustaining that diversity. I then cover a range of topics that includes the poorly understood yet significant ecological processes of organic decay and decomposition; identification of the world's 23 species of Old and New World vultures; the process of convergent evolution; the familial behavior of vultures along with their food-finding abilities, movements, and aspects of their social behavior; and a discussion of the factors that conspire to make vultures one of the most endangered groups of all birds—and how conservations work to protect them.

Although the book generally reflects my own views, it also summarizes the institutional knowledge of many of my colleagues and other professionals in the field who, over the years, have broadened my perspectives considerably, and without whom I would not have been able to describe vultures in this book. They include, but are not limited to, Dean Amadon, David Barber, Marc Bechard, Jim Bednarz, David Bird, Gill Bohrer, Andre Botha, David Brandes, Leslie Brown, Tom Cade, Bill Clark, Mike Collopy, Miguel Ferrer, Allen Fish, Laurie Goodrich, Maricel Grilli-Grana, Frances Hamerstrom, Katie Harrison, Todd Katzner, Roland Kays, Corrine Kendall, Robert Kenward, Lloyd Kiff, Sergio Lambertucci, Yossi Leshem, Mike McGrady, Bernd Meyburg, Ara Monadjem, Joan Morrison, Peter Mundy, Campbell Murn, Juan Jose Negro, Ian Newton, Rob Simmons, Jean-Francois Therrien, Jean-Marc Thiollay, Lindy Thompson, Simon Thomsett, Munir Virani, Rick Watson, Martin Wikelski, David Wilcove, and Reuven Yosef. I also thank my benefactors, Sarkis Acopian and the Acopian family, and my former employer, Hawk Mountain Sanctuary Association. Finally, I thank the fine people at Cornell University Press, including Kitty Liu, Susan Specter, Allegra Martschenko, and Candace Akins.

Also, note that the words raptor, vulture, and buzzard are often confused. Throughout this work I use the term "raptor" to describe all predatory and non-predatory birds of prey, or the hawks, eagles, falcons, vultures, and condors, and the term "vulture" to include all obligate scavenging and largely non-predatory birds of prey, including both vultures and condors. I use the term "buzzard" to describe predatory raptors in the genus *Buteo*.

I conclude with an important disclaimer. In addition to an impressive series of technical papers and books, two monumental monographs have made my work on vultures and condors far easier than otherwise possible. Leslie Brown and Dean Amadon's *Hawks, Eagles and Falcons of the World*, published in 1968, provided me with an essential natural history of the vultures and condors of the world; and Peter Mundy, Duncan Butchart, John Ledger, and Steven Piper's *The Vultures of Africa*, published in 1992, provided an effective model for describing the essential ecology of the world's scavenging birds of prey.

INTRODUCTION

Origins and Evolution of Vultures

The bird student cannot help becoming envious on observing with what accuracy and amazing detail the student of mammals reconstructs the history of that class [with its] fossils . . . Bird bones, being small, brittle, and often pneumatic, are comparatively scarce in fossil collections.
Ernst Mayr (1946)

BEFORE I DESCRIBE the origins and evolution of the birds that ornithologists call vultures, I want to define what vultures are. The world's living vultures represent 23 species of obligate scavenging birds of prey that, unlike raptors, feed principally or exclusively on carcasses they have not killed, but rather have died of other causes. Two of the largest western hemisphere or New World vultures are called *condors* and a few of the larger Old World vultures are called *griffons*. Nevertheless, ornithologists consider them all to be vultures. As do I. Although New World vultures are sometimes commonly referred to as buzzards—most likely because early English settlers in North America confused them with Old World Common Buzzards—New World vultures are not at all closely related to Old World buzzards, which are ancestrally distinct predatory raptors. With that nomenclature in mind, below I discuss the origins and evolution of this trophic assemblage known as vultures.

The world's ecosystems are intrinsically efficient recycling machines. They have to be. Although certain biological taxa, most notably plants, capture and convert inorganic energy into forms that can sustain life, no biological entity, plant or animal, creates its own essential elements (technical terms and jargon are defined in the glossary), which, by definition, make up the indispensable building blocks of life. As a result, the constant recycling of life's limiting essential elements is a critical ecological function of healthy ecosystems. A host of organic decomposers, microbial and invertebrate, serve this function admirably—as do vultures—the world's only group of full-time vertebrate scavengers. Whereas many vertebrates are part-time scavengers, only the vultures do so full time.

But if recycling is essential in ecosystems, why are vultures the only full-time, obligate vertebrate scavengers? Why are there no obligate scavenging fish, amphibians, reptiles, or mammals? The answer lies in a unique suite of what evolutionary scientists call key innovations, specific characteristics that significantly increase an organism's access to ecological resources that enable range expansion, as well as significant population growth, and, in most instances, adaptive radiation and speciation. Examples of key innovations include vertebrate lungs, the paired organs that evolved in a handful of aquatic and semi-aquatic vertebrates, which subsequently enabled the rapid diversification of land-based amphibians, reptiles, birds, and mammals. Innovations also include feathers, the integumentary structures that initially evolved as insulating elements in reptiles, and subsequently as airfoils in birds, enabling their rapid radiation. But what about key innovations in vultures? What innovations do they possess that allow them to be the world's only obligate scavenging vertebrates? A suite of anatomical and behavioral vulturine traits associated with hyper-efficient soaring and gliding flight appears to fit the bill nicely.

The 20th-century American pioneer of animal ecophysiology Knut Schmidt-Nielsen once quipped that flight had "the potential for being the most energetically costly form of animal locomotion." Although Schmidt-Nielsen is right theoretically, natural selection has been shaping avian flight for at least 100 million years, and low-cost soaring and gliding flight rank among its many successes. Today vultures use soaring and gliding flight to achieve full-time obligate scavenging in two ways. First, such flight elevates vultures above the landscape so they can see far enough to scan effectively both for carcasses and other vultures, and second, it allows them to travel long distances at minimal energetic cost while doing so. That vultures frequently rely on cues from one another to help them discover carcasses also increases the chances of successfully finding their next meal. Hence, social behavior serves as a key innovation in the development of obligate scavenging. As do their relatively large body mass and crop size, which allow them to feed sufficiently at a single meal to store food enough to sustain them for several days. Together, this suite of characteristics has enabled three distinct vulturine lineages to achieve something unique among vertebrates: an obligate scavenging lifestyle.

How did this unique avian trophic guild evolve? A somewhat far-fetched energetics model developed by two Scottish biologists helps explain why hyper-efficient soaring and gliding flight is critical. Graeme Ruxton and David Houston developed their energetic model to evaluate the potential for obligate scavenging in a ground-based vertebrate that few might have anticipated: the long-extinct, ferocious, meat-eating, and up to 15 metric-ton dinosaur *Tyrannosaurus rex*, a species the two concluded would have been capable of full-time, obligate scavenging in the Serengeti savannas of East Africa, at least without obligate avian scavengers. As far-fetched as it at first might seem, Ruxton and Houston's case is convincing in that East Africa's ungulate populations and the carcasses they provided would have been numerous enough to support the nutritional needs of a small population of these dinosaurs. However, only if *Tyrannosaurus* met a key requirement for successful obligate scavenging: hyper-efficient locomotion. *Tyrannosaurus* would have had to reduce the cost of movements to a bare minimum, something Ruxton and Houston suggest a lumbering, slow-moving version of the gigantic reptile could have accomplished. Although somewhat fanciful, their model demonstrates that low- to no-cost movement ecology is key to developing an obligate scavenging lifestyle, something vultures have managed to achieve by means of efficient soaring and gliding flight.

Although previous experts had proposed the *Tyrannosaurus* was a facultative scavenger more than a century ago, American paleontologist Jack Horner built a stronger case for their obligate scavenging lifestyle in 1994, when he posited that the species' poor eyesight, "puny" forelimbs, grinding rather than piercing teeth, and ponderous gait combined to so compromise potential predatory behavior as to make obligatory scavenging a foregone conclusion. Though others have questioned Horner's line of reasoning, the species' towering upright adult stature of more than 5 meters certainly would have permitted it to scan larger areas of surrounding open habitats for large carcasses, which is another important scavenging trait. Equally important was its relatively high abundance (as compared to large co-occurring herbivorous dinosaurs). A multi-species census in fossil beds in the Hell Creek Formation of northeastern Montana ranked it as the second most abundant of all dinosaurs with population densities rivaling those of hyenas in the Serengeti of East Africa, which supports the likelihood of *Tyrannosaurus rex* being largely, if not entirely, a scavenging versus a predatory beast.

But enough about the *why* and *how* of vertebrate scavenger evolution; on to the *where* and *when*.

ANCIENT AVIAN LINEAGES

The fossil remains of vultures, like those of other large-bodied and large-boned birds, are reasonably well represented in the avian fossil record. Based on that record, avian paleontologists now recognize three genetic lineages of obligate scavenging raptors that fall into two avian families: **Cathartidae**, represented by seven living species of New World vultures, including the Turkey Vulture, Lesser Yellow-headed Vulture, Greater Yellow-headed Vulture, Black Vulture, King Vulture, California Condor, and Andean Condor; and **Accipitridae**, represented by 16 living species of Old World vultures in two distinct subfamilies, the **Gypaetinae**, including Palm-nut Vulture, Egyptian Vulture, and Bearded Vulture, and the **Aegypiinae**, including the Red-headed Vulture, Cinereous Vulture, Lappet-faced Vulture, Hooded Vulture, Indian Vulture (aka Long-billed Vulture), Slender-billed Vulture, Cape Vulture, Griffon Vulture, Rüppell's Vulture, White-backed Vulture, Himalayan Vulture, and White-rumped Vulture. A fourth lineage, potentially obligate, but more likely facultatively scavenging, **Teratornithidae**, is detailed in Box 0.1.

Box 0.1 Teratorns: A Third Group of Obligate Avian Scavengers

Teratorns, an extinct New World lineage of large scavenging and predatory birds of prey, represent one of the most remarkable assemblages of avian fossils. Members of the extinct avian family Teratornithidae, these so-called wonder birds, included the largest of all flying birds.

The family Teratornithidae was introduced to science in 1909 when the University of California ornithologist and paleontologist Loye Miller who was in the process of retrieving and studying fossil remains from tar pits in Los Angeles, California, described one teratorn as follows:

> Among many interesting forms of vertebrates taken from the Quaternary asphalt of the Rancho La Brea beds in Southern California

> have appeared several specimens of a very large bird which show marked divergence from recent forms [so] as to necessitate the establishment of a new genus.

Rancho La Brea had already yielded fossilized bones of scavenging raptors, including those of both California Condors and Turkey Vultures, but teratorns were distinct from these raptors. Bigger, more robust, and with anatomical hints of New World vulture "affinities," the group now includes at least four extinct forms. The most widely recognized is Miller's original find, *Teratornis merriami*, a bulked-up, Arnold Schwarzenegger–like version of the California Condor that stood about 30 in (75 cm) tall and weighed an estimated 33 lb (15 kg). As in other teratorns, *merriami* had an eagle-like bill, suggesting that it could guzzle smaller prey whole as well as hastily devour larger pieces of more massive prey in much the same way that hugely beaked bands of Steller's Sea Eagles do today. It also possessed what Miller described as a "high bridge of the nose" and it lacked a nasal septum, features that link it to extant New World vultures. The La Brea tar pits would have been excellent habitat for scavengers like teratorns as its glossy surface probably attracted mammals and other vertebrates, numbers of which would have become mired in the shimmering puddles of liquid asphalt they had mistaken for watering holes in the dry California landscape.

Although Miller was quick to point out of the many differences between *Teratornis merriami* and extant New World Vultures, he was especially taken by the specimen's head region, which displayed a "striking similarity" to that of the California Condor, and the evident "preponderance of Cathartid affinities."

Subsequently discovered fossil teratorns include a smaller version of the La Brea specimen from southern Brazil, but the real subfamily show-stopper is *Argentavis magnificens* from more southern South America. Dating from the late Miocene 5 to 8 million years ago, this extinct teratorn stood 5 ft (1.5 m) tall, had a wingspan of more than 23 ft (7 m), and weighed more than 150 lb (70 kg). Specialists indicate that although it, too, was capable of soaring flight, strong Patagonian westerlies would have been necessary for takeoffs as individuals spread their enormous wings.

All fossil teratorns feature stout, fast-scampering legs, and both living prey and fresh carcasses, most likely, would have nourished all of them.

The group's disappearance at the end of the Pleistocene 11,000 years ago coincided with the large-scale collapse of New World ungulate megafauna, either directly, due to the disappearance of its food base, or because of climate change. The coincidental disappearance of teratorns and North American megafauna supports the idea that teratorns fed mainly on large carcasses and may have been episodically predatory as well.

German museum curator and ornithologist Hans Gadow was the first to note that so-called New World and Old World Vultures differed from each other, not only in their global distributions but also in several notable anatomical characteristics, including the perforated nares and double-notched sternum in New World but not in Old World vultures. Numerous analyses have confirmed these and other anatomical and behavioral differences, and the ancestral history of New World versus Old World vultures remains essentially settled. The once widely held belief that New World Vultures were more closely related to storks than to Old World Vultures and other birds of prey has largely been refuted. Assuming this to be realistic, the avian families Cathartidae and Accipitridae represent one of the best examples of what evolutionary biologists call *convergent evolution*, the independent evolution of similar characteristics in species or groups of species from different lineages resulting from ecological rather than evolutionary similarities (Box 0.2).

Box 0.2 Convergent Evolution

Convergent evolution, "the independent evolution of structural or functional similarity in two distantly-related phylogenetically distinct lineages not based on common ancestry," provides evolutionary biologists with one of the most convincing arguments for the power of natural selection.

Widely recognized examples of convergent evolution in vertebrates include ecological and anatomical similarities between the phylogenetically distinct North American gray wolf and the southern hemisphere Tasmanian wolf; African sunbirds and American hummingbirds; sharks and dolphins;

and New World and Old World vultures. Although the latter two lineages are not particularly closely related, the two exhibit many similarities, several of which include:

Anatomical similarities

1. *Bare or minimally feathered heads.* Most species in both groups feature completely bare or minimally feathered heads, most likely because it makes it easier for individuals to keep their heads clean when feeding on and inside of carcasses.
2. *Oversized wings.* Species in both groups have large, oversized wings that result in lighter wing loading, a feature that facilitates sustained soaring flight and permits low-cost searching for carcasses.
3. *Relatively large body masses.* In both groups, body mass ranges from slightly less than 2.2 lb (1 kg) to more than 33 lb (15 kg). Large body mass permits alternating episodes of gorging and prolonged periods of non-feeding when carcasses are consistently available.
4. *Distensible and visible crops.* Representatives of both groups have visibly distensible crops that signal recent feeding, which serve to share useful information regarding successful feeding among individuals at communal roosts.

Physiological similarities

5. *Controlled torpor.* Representatives of both groups engage in controlled nocturnal torpor (a daily reduction in core body temperature) to save energy.
6. *Low basal metabolic rate.* Representatives of both groups exhibit low basal metabolic rates for their size.

Behavioral and ecological similarities

7. *Communal feeding.* This occurs particularly at large carcasses given that food availability exceeds the storage capacity of individual birds.

8. *Communal roosting.* This enhances the likelihood of information exchange among departing individuals regarding the location of carcasses.
9. *Social networking when searching for food.* Individuals in both groups often rely on cues from conspecifics to help locate carcasses.
10. *Niche partitioning among species in feeding behavior in both groups.* Species of both groups demonstrate species-specific behavioral and anatomical separation in feeding types in the shapes of their skulls, beaks, and mandibular dimensions that result in distinctive feeding types, which reduces competition at carcasses.

Plumage convergence between species

11. *King Vultures and Egyptian Vultures.* Although these two species differ considerably in the structure and color of their head and neck, as well as in the color of their tail, adults of both have distinctive, largely white or cream-white body feathers, and upper and lower wing coverts, with black primary and secondaries.
12. *Turkey Vultures and Red-headed Vultures.* Although the two species differ considerably in the structure of their head and neck, adults in both species have largely bare-skinned red or reddish heads and largely black or blackish body feathers and upper and lower wing coverts, and silvery or whitish gray bases to their wing feathers.

None of the convergences highlighted above negate the fact that New Word and Old World vultures differ in many significant ways. New World vultures, for example, have proportionately and notably smaller toes than do Old World vultures, and three species of New World vultures "smell for" as well as "look for" carcasses. Finally, New World vultures are decidedly less specific in habitat use.

Although initial assessments of the origins of Cathartidae and Aegypiinae-Gypaetinae placed the first in the New World and the latter two in the Old World, more recent fossil evidence suggests that the New World family Cathartidae most likely originated in the Old World Eocene more than

50 million years ago, and that both Old World subfamilies occurred in the New World Miocene more than 15 million years ago. Regardless of what we now know about their apparently convoluted geographic origins, anatomical differences between New and Old World vultures are such that Loye Miller, the California museum curator who unearthed the first fossils of Old World vultures in California, USA, in the early 20th century, delayed announcing his finding for several years because he thought his colleagues might not believe him. Also, although the terms New World and Old Word vultures remain widely accepted, it is important to keep in mind that they refer only to current species distributions of the two groups and not their geographical origins.

New World Vultures. New World, Cathartid, vultures first appear in the Old World fossil record dating from Lower Eocene and Oligocene of Europe, as far back as 50 million years ago, and in the lower Oligocene of Colorado 30–40 million years ago. Intriguingly, two of the earliest Cathartid fossils, *Plesiocathartes* and *Diatropornis*, were relatively small "raptorial-like" birds, leading some to suggest they were not specialized carrion feeders. The group's appearance in Old World Eocene sediments suggests Cathartidae most likely originated and radiated in warmer parts of Eurasia sometime in the late Cretaceous or early Eocene, before "emigrating" to and thereafter radiating in the western hemisphere, where numerous Pleistocene fossils occur. Indeed, all seven living Cathartidae occur in the Pleistocene as well, as do several close relatives including a slightly larger version of the California Condor; a more massive, but shorter and bulkier-footed version of the Black Vulture; and a King Vulture "type" that was intermediate in size between its living counterpart and the Andean Condor.

As yet unexplained is the family's ancestral extirpation in Europe. Whether that elimination resulted from changing climates, competition with evolving Old World vultures, or something else entirely, is largely undiscussed in the literature.

Old World Vultures. As mentioned above, Old World, or Accipitrid, vultures evolved into two separate and distinct subfamilies, the Gypaetinae and the Aegypiinae. The former, which is more ancient as well as more ecologically diverse, is particularly fascinating in that its lengthy history provides useful insight into origins of the vulturine lifestyle, at least within the subfamily. Gypaetinae currently includes the Palm-nut Vulture, a species once thought to be closely related to sea eagles, along with the Egyptian Vulture, the Bearded Vulture, and curiously enough, the non-vulturine Madagascar Serpent Eagle, all of which differ considerably in diet. The Palm-nut Vulture is mostly vegetarian and secondarily predatory, as well as carrion eating. The Egyptian Vulture is an opportunistic obligate scavenger whose eclectic diet includes insects, human rubbish, scraps from larger carrion, vulnerable young and injured small vertebrates, and ostrich and other large-bird eggs. The Bearded Vulture is a dietary specialist that feeds on large bones, which it breaks into small pieces, and on both non-bone carrion and living prey. The Madagascar Serpent Eagle is a predatory raptor that largely takes lizards and tree frogs. (Note: I have described the Palm-nut Vulture in this book because of its current common name, ancestry, and part-time carrion-eating behavior, but I have not described the Madagascar Serpent Eagle both because of its name and because, ecologically, it is not very vulture-like.) Although ecologically and somewhat genetically divergent from other vultures, these four species are related, genetically, more closely to each other than to other raptors including, most notably, the somewhat distantly related Aegypiinae. Both molecular and anatomical data indicate that the subfamily Gypaetinae is closely related to the Perninae, a predatory subfamily that includes honey-buzzards and several other medium-sized, broad-winged, and largely tropical raptors.

The more speciose Aegypiinae includes all eight living species of *Gyps* vultures, a group of large, sandy to brownish, long-necked vultures of southern Europe, Africa, and southern and central Asia. *Gyps* are believed to have diversified in response to a global grassland transition that created vast regions of megafaunal-filled landscapes during the early Miocene and late Pliocene epochs 23 to 2.5 million years ago, together with four larger, thicker-billed, and darker Old World species, including the Red-headed Vulture, White-headed Vulture, Lappet-faced Vulture, and Cinereous Vulture. The exact relationship of the much smaller and thinner-billed Hooded Vulture within the subfamily remains unclear. Ancestrally, the Aegypiinae is closely related to Accipitrid subfamily Circaetinae, a largely predatory taxon that includes snake and serpent eagles, as well as the largely scavenging Bateleur.

Old World fossil vultures include a questionable form from the late Miocene–early Pliocene of South Africa, *Paleohierax*, that appears similar to the Palm-nut Vulture. The earliest known western hemisphere specimens of Old World vultures include a *Palaeoborus* from the lower

Miocene of South Dakota and a second *Palaeoborus* species from the middle Miocene of Nebraska. More recent fossils from the New World indicate an arguably unexpected diverse species assemblage unrivaled by Old World fossils. Many New World Pleistocene Aegypiinae fossils are contemporaneous with numerous Cathartid forms, and the extent to which competition between the two clades played a role in the former's demise remains unknown.

In sum, it is fair to say that the fossil history of vultures, although intriguing, is also frustrating, in part because it cannot be studied in the field with living organisms. That said, the ecology of the living descendants of these fossil forms are the subjects of the seven chapters that follow.

SYNTHESIS AND CONCLUSIONS

1. The world's 23 species of vultures are the world's only obligate scavenging vertebrates.
2. The fossil remains of vultures, like those of other large-boned birds, are relatively well represented in the avian fossil record.
3. Several key innovations, including hyper-efficient, low-cost soaring, and gliding searching flight, along with a reliance on visual cues from the flight behavior of other vultures to discover carcasses, enable vultures to achieve their obligate scavenging lifestyle.
4. The Neogene, a geologic period that spanned approximately 20 million years from 2.5 to 23 million years ago, saw the development of seasonal grasslands and large populations of ungulates that fostered the adaptive radiation of many obligate scavenging birds, including many vultures.
5. Three living ancient lineages, the Cathartidae, Gypaetinae, and Aegypiinae, are ancestral to all living vultures.
6. One ancient extinct genetic lineage, the Teratornithidae, includes a number of extinct Teratorns that most likely included obligate scavenging birds.
7. What we now call Old World Vultures may have evolved in the New World, and those we call New World Vultures most likely evolved in the Old World.

1 ESSENTIAL ECOLOGY OF SCAVENGERS

In every community we should find herbivorous,
carnivorous, and scavenging animals.
Charles Elton (1947)

THE PIONEERING ENGLISH ECOLOGIST CHARLES ELTON was spot-on in his assessment of the importance of avian scavengers. But although carrion consumption is an essential aspect of animal ecology, the consumption of animal carcasses has been an academic backwater until recently. This scarcity of information has much to do with humanity's long-standing aversion to the dead, a deep-seated adaptation to the possibility of human and livestock contagions.

Mounting evidence suggests that in most landscapes, decomposition and scavenging both remove and effectively recycle significant organic material, and as such are essential ecological processes worthy of study. Across the savannas of East Africa, an ecosystem in which these phenomena have been studied in considerable detail, scavengers and decomposers consume and recycle 70% of all of the large ungulate biomass, whereas lions and other predators remove only 30%. In the Arctic near Barrow, Alaska, scavengers, including grizzly bears, wolves, and Northern Ravens, removed half of all specifically placed large-mammal carcasses in summer; and foxes, least weasels, and lemmings scavenged 99% of all subnivean lemming carcasses in winter. In and around residential Oxford, England, where predators account for 60% of small-mammal deaths, scavengers and decomposers consume 40% of all carcasses annually. Examples such as these suggest that single-celled decomposers and multicelled scavengers, including vultures, represent two of the most important groups of organisms in ecosystems. Appreciating their importance is key to our fully understanding how both natural and human-dominated landscapes function.

ECOLOGICAL FUNCTIONS OF SCAVENGING AND DECOMPOSITION

Understanding the ecological roles of decomposers and scavengers requires understanding differences in the functions and attributes of these organisms in ecosystems. A brief primer in ecosystem science and Newtonian thermodynamics is useful. As many learn in ecology 101, ecosystems are made up of both producers and consumers. Producers, or autotrophs, include plants and microorganisms that make their own food by transforming solar energy into chemical energy in the form of organic, carbon-based substances. Plants do so by means of the intricate biochemical process of carbon fixation known as photosynthesis, and several autotropic microorganisms do the same through various processes in chemosynthesis. Consumers, most of which are heterotrophs, then use this organic material to fuel their metabolic, behavioral, and reproductive activities. The transfer of food energy from producers through one or more animal consumers is called a predatory food chain. Ecological links in food chains are determined by how many types of animals are consumed or preyed upon by other typically larger predators. In many cases, animals—and some of their organic matter—are consumed by smaller organisms called parasites. In other cases—and this is where it gets interesting—the carcasses of dead animals that have been either preyed upon by larger animals or parasitized by smaller organisms, or have died by metabolic processes attributable to misfortune or accident, are consumed by decomposers, scavengers, or both.

Overall, decomposers and scavengers function similarly in ecosystems, the principal differences being their respective size and where and how their "feeding" occurs. Decomposers are mainly single-celled, saprophytic or saprozoic organisms, including bacteria, that feed extracellularly on dead material and subsequently absorb their food in place. Scavengers, most of which are multicellular organisms, including invertebrates and their young—think maggots—and vertebrates including vultures that feed on the fragments of carcasses and subsequently ingest them, do so either on-site or elsewhere. Energy transfer is key in all of this. The transfer of energy, or the "ability to do work," from one form to another is governed by Newtonian Laws of Thermodynamics. Sir Isaac Newton's First Law states that although energy can be transformed from one form to another, it can never be created or destroyed, whereas his Second Law states that energy transformations always involve the degradation of energy from more concentrated to more dispersed forms. All energy transformations, ecological and otherwise, are "leaky processes" that result in the dispersion of energy from a more valuable concentrated form to a less concentrated and, therefore, less valuable form that is less useful in doing work, ecologically and otherwise. Because of the laws of thermodynamics, "pyramids of energy" are said to occur in ecological food chains, with less useful energy being available at every subsequent link in the chain. This, in turn, is why decomposers and scavengers are essential in ecosystems. Although decomposers and scavengers, like other organisms, obey the laws of thermodynamics, their ecological actions as elemental recyclers serve to recapture and supply the materials or essential ingredients that other ecological actors, including both plants and consumers, need to maintain the energetic processes of primary and secondary productivity that enable the long-term continuity of ecological systems. Thus, whereas many of my colleagues in conservation education typically describe vultures, and for that matter other vertebrate scavengers, as nature's "sanitation engineers," these organisms in fact are much more than that. In addition to keeping ecosystems clean and reducing the likelihood of both livestock and human diseases, decomposers and scavengers perform essential work as elemental recyclers that serve to reconstitute the raw materials that propagate and perpetuate long-term ecosystem function.

Decomposers versus Scavengers

Decomposers and scavengers play critical roles in speeding material recycling in ecological systems. Strictly speaking, decomposers do their recycling mainly "out of body," or extra-cellularly, and on-site. Scavengers, including vultures, remove organic material from decaying carcasses for subsequent digestion and disintegration elsewhere. The dependence of decomposers and scavengers upon similar, if not identical, resources—dead and dying organic matter—means that they compete with one another for their food. This ecological reality functions to hasten their feeding rates. Estimates suggest that without the gluttonous and frenzied feeding behavior of these organisms, decomposition would slow tenfold. Vultures compete not only with other avian scavengers but also with many additional types of organisms. Although the carcasses themselves are defenseless, the degree of both intra- and interspecies competition within these feeding communities can be brutal, both physically and chemically.

Hundreds, if not thousands, of species of decomposers settle on, within, and around both the carcasses and soon-to-be-carcasses of dead and dying animals, proceeding rapidly to mineralize organic waste while producing food for themselves and, in many cases, producing a suite of chemical agents that can affect the success of their competitors. As a result, decomposer metabolism at carcasses often proceeds at rates that heat the carcasses of disintegrating warm-blooded animals to temperatures that approach or exceed what they were prior to death.

Competition by decomposers at carcasses takes countless forms, many of which are largely unappreciated by avian biologists. Typically, the struggle begins as inter- and intra-species chemical warfare, as physical warfare is out of the question given the oftentimes enormous size differences among the combatants. In such situations, microbes quickly fabricate off-putting and often toxic metabolites to fend off other microbes and larger competitors. The creation of alcohol and other metabolites by many decomposers can inebriate, intoxicate, and sicken vertebrate scavengers, making them vulnerable to avian and mammalian predators. As a result, livestock, rodents, and herbaceous scavengers are known to avoid moldy grain and rotten fruit that can cause intestinal discomfort. Indeed, the well-known antibiotic fungal metabolite, penicillin, most likely evolved to forestall bacterial growth in "moldy" organic matter. The objectionable nature of over-ripe fruits and vegetables suggests the widespread nature of this type of chemical warfare among decomposers. That disease-causing *Salmonella*, *Staphylococcus*, and *Clostridia* produce toxins in the tissues they infest supports this line of reasoning.

The American evolutionary ecologist Daniel Janzen authored a paper intriguingly titled "Why fruit rots, seeds mold, and meat spoils" in 1977. In that paper, Janzen laid out a cogent argument that suggested that persistent interspecies competition between microbes and large organisms would create strong selective pressure on microbes to chemically render the seeds, fruits, and meats they were consuming inedible to vertebrates, leading scavengers to develop the ability to "ignore or mask objectionable flavors or odors that are also warnings and [thereafter] detoxify or otherwise avoid" them as much as possible.

In addition, many vertebrate scavengers host detoxifying symbiotic microbes to cope with rotting meat. Microbes in the lower gut of these individuals often create antibiotics that selectively clear this region of competitive microbes. A recent investigation of the microbiomes of 50 individuals of two species of New World vultures demonstrated the remarkable "conservancy" of a small number of types of gut flora dominated by the microbial genera *Clostridia* and *Fusobacteria*, which consist of species of common soil bacteria that typically are toxic to other vertebrates (for example, tetanus). Intriguingly, alligators, which also scavenge carrion, have similar microbial communities. The presence of *Clostridia* and *Fusobacteria* simply may be due to the organisms' ability to outcompete other bacteria in the lower digestive tracts of New World vultures. Yet, it may be fostered by scavenging birds and reptiles for their own benefit in the breakdown and use of other carrion bacteria for their food value, while escaping the possible negative effects of related toxins. Several New World vultures possess antibodies for botulinum, leading researchers to speculate that they may be able to tolerate such toxins.

Another likely symbiotic relationship between vultures and their microbes involves *urohidrosis*, the socially unacceptable phenomenon of defecating on one's feet and toes. Several New World vultures do this predictably and routinely. Although the behavior has long been attributed to evaporative cooling on particularly warm summer days, individual vultures do so in winter, suggesting the likelihood of an additional, antiseptic, function as individuals slog through putrefying carcasses. Whether this is true remains to be tested.

Finally, if carcasses become chemically less valuable to vultures over time, quickly locating and consuming carrion would be selected for.

Facultative versus Obligate Vertebrate Scavengers

Two types of vertebrate scavengers occur. Most vertebrate scavengers are *facultative* scavengers, species that sometimes but not always feed on carrion. A distinct minority are *obligate* scavengers, species that always, or almost always, feed on carrion. Because facultative scavengers are predatory as well as scavenging vertebrates, they are notably less specialized anatomically, behaviorally, and nutritionally for carrion finding and feeding, and they display fewer adaptations to the scavenging-only lifestyle than do obligate scavengers. Only several dozen avian scavengers are obligate, including vultures, several caracaras, and two species of giant petrels. No living mammalian scavengers are in the obligate group.

Facultative Scavengers. Facultative scavenging is phylogenetically widespread. Almost all predatory organisms, including the overwhelming majority of raptors and other carnivorous vertebrates, are facultative scavengers. The degree to which scavenging occurs varies opportunistically among species of facultative scavengers, as well as both seasonally and geographically within species, and, in some instances, individually. Jaguars, lions, and grizzly bears, for example, do not forego opportunities to feed on carcasses while searching for living prey, as do many eagles, hawks, and falcons. In fact, many predators routinely scavenge, including wolves, wild dogs, foxes, jackals, hyenas, bobcats, and lynx. Although fewer reptiles than mammals and birds routinely scavenge carrion, alligators and many lizards do, including Komodo dragons and other monitor lizards, all of which use their sense of smell to locate dead animals. Many predators scavenge by displacing less weaponized obligate scavengers and other predators from recently killed prey, whereas others feed on animals that have died from injury, disease, malnutrition, and other causes. Seasonality is important, with many carnivores switching from predation to scavenging during lean times of the year when live prey is less readily available. In some instances, predators undertake seasonal shifts in their feeding niches. In winter in North America, for example, wolverines typically search for and scavenge the carcasses of winter-storm-killed animals, whereas in summer they return to preying upon newborn reindeer and caribou. In East Africa, relationships among predator-scavengers are complex with predictable role reversals among species, for example, lions routinely scavenge prey remains left by hyenas and vice versa. Another complexity occurs when both species prey upon each other depending upon the circumstances, for example, the relative health of the individuals involved, the relative densities of their populations, and shifts in their group sizes and social structures. In more open habitats with limited cover, hyenas actually benefit lions when sustained pursuit predation is more profitable, whereas lions and leopards can benefit hyenas in brushier areas where successful lengthy pursuit can be more difficult for the latter.

Obligate Scavengers. Sixteen species of Old World vultures and seven species of New World vultures are generally recognized as the only truly obligate vertebrate scavengers, that is, the species that feed all but entirely on carrion and that require carrion for their livelihoods. (The California Condor, Andean Condor, and Eurasian Griffon are included here as "vultures" as well, as they are phylogenetically and ecologically part of this trophic assemblage.) All 23 species have large wingspans and light wing loadings that allow them to soar at low cost over large areas in search of episodically and ephemerally available dead animals (or carrion). All possess keen eyesight and several possess a keen sense of smell to help them locate carrion. All have reduced feathering around their heads that might otherwise cake their heads with putrefying material, at least occasionally. And all use social networks to help locate carrion more efficiently. This suite of somewhat specific adaptations allows them to outcompete other vertebrate scavengers that, in addition to feeding on carrion, depend on live prey to supplement their carrion diet.

SYNTHESIS AND CONCLUSIONS

1. Because of humanity's long-standing aversion to the dead, microbial decomposition and carrion consumption by scavengers remain largely "neglected" aspects of healthy ecosystems despite their essential ecosystem function.
2. Decomposers are single-celled, saprophytic or saprozoic organisms, including bacteria, that feed extracellularly on dead material on-site. Scavengers are multicellular organisms, including both invertebrates and vertebrates, that feed on the fragments of carcasses, either on-site or elsewhere.
3. The ecological actions of decomposers and scavengers as elemental recyclers serve to recapture and supply the raw materials that other ecological entities, including plants and consumers, need to maintain the energetic processes of primary and secondary productivity necessary to sustain the long-term continuity of ecological systems.
4. Many microbial decomposers engage in chemical warfare by fabricating off-putting and, at times, toxic metabolites to ward off

other microbes and larger competitors. This chemical process is why plants mold, fruits rot, and meats spoil.

5. Urohidrosis, the habit of defecating on one's feet and toes practiced by several New World Vultures, which has long been attributed to evaporative cooling on hot days, also occurs in winter, suggesting the possibility of an antiseptic function against microbial infection.
6. Although many carnivorous vertebrates are facultative scavengers that sometimes feed on carrion, the world's vultures are the only truly obligate vertebrate scavengers, species that feed almost entirely on carrion and that require carrion for their livelihoods.
7. All 23 species of vultures have large wingspans and light wing loadings that allow them to soar at low cost over large areas in search of episodically and ephemerally available dead animals, also known as carrion.

2 SPECIES DESCRIPTIONS AND LIFE HISTORIES

Among extant vertebrates, only 23 species of vultures are obligate scavengers.

G. D. Ruxton and D. C. Houston (2004)

ALTHOUGH WIDELY DISTRIBUTED, obligate scavenging birds of prey are not particularly diverse. Only 7 species occur in the New World and only 16 species occur in the Old World. Altogether, these 23 species make up fewer than 1% of all birds of prey and fewer than three-thousandths of all birds globally. Vultures, however, are widespread geographically. Their populations are most numerous and their diversity highest in tropical and subtropical regions. Ranging from 55° North to 55° South in the New World, and from 51° North to 34° South in the Old World, one or more species occur on all continents except Australia and Antarctica. Africa has 11 species; Asia has 9; South America, 6; Europe, 5; and North America, 3.

Aside from vagrant Cinereous Vultures in the Philippines, modern vultures are largely absent from the South Pacific, most likely because of a lack of large mammal carcasses in that area. That said, remains of a large Old World vulture, similar if not identical to Africa's White-headed Vulture, are known from late Pleistocene sediments on the island of Flores in the South Pacific. Now, however, except for the Mediterranean islands, only two species, Turkey Vultures and, to a lesser extent, Egyptian Vultures—both of which are known to make short water crossings—are not particularly widespread on islands. Even so, the combined distributions of vultures include well over half of the world's continental land masses, and, as indicated earlier, their scavenging behavior makes them functionally significant components of the ecological landscapes they occupy.

Substantial differences in both the distributions and the abundances of individual species occur, and together with their different phylogenies and their anatomical and behavioral characteristics, all species distinguish themselves as unique biological entities. Geographic populations *within* species, too, often differ behaviorally and sometimes even anatomically, as do individuals *within* populations. Such biological variation provides the raw material for evolution by natural selection, and several currently recognized "species," including both Turkey Vultures and Hooded Vultures, are believed to consist of two or more cryptic species.

Below I describe and detail the life histories of these species. The common and scientific names I use are the widely held preferences of the scientific and conservation communities of 2019 and primarily follow those of Gill and Wright's *Birds of the World: Recommended English Names.* (Order of appearance in the Life Histories also follows Gill and Wright 2006.) Their estimated population sizes are, with few exceptions, from BirdLife International 2019. The descriptions that follow differ in detail and length given what is currently known for each.

Throughout this chapter I focus on two principal and equally significant messages. First, marked species differences are found within this trophic guild, and such differences permit vultures to more effectively and fully use carrion as a nutritional resource in ways that allow them to maintain far larger populations than would be possible otherwise. Second, although the geographical and ecological rules of nutritional ecology set forth elsewhere in this book largely hold, species diversity within this guild allows these rules to be applied successfully in a variety of ways.

NEW WORLD VULTURES

Black Vulture *Coragyps atratus*

Spanish: Zopilote Negro, Aura Negro
German: Rabengeier
French: Urubu Noir
aka: American Black Vulture, Gallinazo, Carrion Crow (obsolete)

Taxonomy: Family Cathartidae

Size: Length 26 in (56–74 cm); wingspan 58 in (150 cm); 3.0–5.0 lb (1400–2200 g)

Estimated world population: >>10,000,000; increasing
Movements: Largely sedentary except near latitudinal limits of its range; sometimes nomadic
Social ecology: Individual nesting; communal roosting; individual and group food finding

With a species range of more than 17 million mi^2 (44 million km^2), the Black Vulture ranks second to the Turkey Vulture as the most widely distributed vulture in the world. Decidedly more abundant than the latter in many portions of its tropical and subtropical range, the Black Vulture is, by far, the most common bird of prey in many parts of the Americas and is considered by many to be the most common raptor globally, with one estimate suggesting as many as 20 million individuals.

The species is a relatively small, blackish, scavenging bird of prey with a body mass that varies depending on subspecies from <2 lb (<1 kg) in the tropics, to as much as 5 lb (2.2 kg) in the northern temperate zones. Black Vultures soar both high and low while searching for both small and large carcasses, as well as for human refuse. The species occasionally takes vulnerable and largely helpless living prey. Black Vultures favor riparian and naturally humid areas more so than do Turkey Vultures, and they are often less common than the latter in arid areas. Gregarious and often commensal, Black Vultures do well in human-dominated landscapes throughout their range and are likely more common today than at any time previously.

The species is highly social and frequently aggressive at carcasses; as such, it is more similar behaviorally to Old World *Gyps* species than to other New World vultures. Black Vultures occur in the fossil record of both Pleistocene North and South America.

Charles Darwin was the first to mention that Black Vultures had a more northerly distribution—that is to say, a more tropical distribution—in South America at the time of European conquest, and that its distributional growth "followed [human] inhabitants from more northern districts" to many parts of its expanding distributional range in South America. By the early 1800s, Black Vultures were abundant inhabitants of many cities in both North and South America, where it was seen "sauntering about" as "tame as poultry" and "completely domesticated." Although many early observers considered the "carrion crow" to be filthy and possessing fearful voracious habits, Black Vultures were widely recognized as useful elements in removing organic waste at the time and were protected by law in certain municipalities, including coastal Charleston, South Carolina, USA. The founder of American Ornithology, Alexander Wilson, had this to say about the unreserved nature of the species in 1840:

> Went out to Hampstead [a neighborhood on the outskirts of Charleston, South Carolina] this afternoon. A horse had dropped down [dead] on the street . . . and was skinned. The ground, for a hundred yards around it, was black with carrion crows [Black Vultures]; many sat on the tops of sheds, fences, and houses within sight, sixty or eighty on the opposite side of a small run. I counted, at one time, two hundred and thirty-seven, but I believe there were more, besides several in the air over my head, and at a distance.
>
> I ventured cautiously within 30 yards of the carcass. . . . seeing them take no notice I ventured nearer, till I was within 10 yards, and sat down on the bank. Sometimes I observed them stretching their necks along the ground, as if to press the food [therein] downwards.

As human sanitation improved over the course of the 19th and early 20th centuries, the species' value in carcass disposal declined. Human fear of the birds increased during this timeframe given their tendency to pursue and sometimes kill vulnerable young poultry, piglets, and lambs as well as concern over their spreading both livestock and human diseases. The species that was once viewed as a human asset was reclassified as bothersome and potentially dangerous, and was persecuted in many parts of its range. Current legal protection in many parts of its distributional range has decreased persecution overall. Many populations have since increased, and the northern limits of its range have since expanded. Black Vultures now breed in the northeastern United States and southeastern-most Canada, where they sometimes overwinter.

The Black Vulture was described as *Vultur atratus* by J. M. Bechstein in 1793, after being named by the American naturalist William Bartram, based on a specimen collected along the St. John's River in northern Florida. The genus name *Coragyps* is from the Greek *korax*, for "raven," and *gyps*, for "vulture"; the species name *atratus* is from the Latin for "clothed in black."

Subspecies. Three subspecies of Black Vultures are recognized, all of which are identifiable in the field. From north to south the three subspecies include the nominate *atratus*, which ranges from eastern Canada in North America to the subtropics of Mexico in Central America; *brasiliensis*, a smaller bird with larger and bolder white markings on the outer underwings, which ranges from tropical Mexico to coastal Peru and west to the lowlands of Bolivia and Brazil; and *foetens*, which is about the same size as *atratus* but has less prominent white wing patches and ranges from the northern Andes south into the lowlands of Chile, Paraguay, Uruguay, and Argentina south to the Rio Negro.

As is true of the Turkey Vulture, the Black Vulture has been studied more in the United States than elsewhere in its range, particularly with regard to social behavior.

Breeding biology. The species does not build a nest, but rather lays typical two-egg clutches in dark recesses, on base soil, stones and gravel, fallen timber, and on the floors of abandoned and little-used buildings. Considerable variation in nest-site preferences occurs, even over short distances. In relatively open habitat farmland country in eastern Pennsylvania, USA, for example, most nests sites are in small, abandoned farm buildings, often within 30–60 ft (10–20 m) of more active buildings and residences. Less than 85 mi (150 km) away in the Gettysburg National Military Park, most nests are in rock crevices, which shows a "significant preference for nest sites in roadless forested areas with no buildings." Overall, the species frequently reuses successful nest site in successive years, suggesting "local nest-site cultures" based on previous nesting success.

Detailed studies involving multiple nests and multiple years of the nesting biology of individuals suggest that nesting sites are limited. One study involving more than 60 nests in farmland in central North Carolina, USA, where Black Vultures had nested at least since the late 1700s, was undertaken in response to suggested regional declines in the 1970s. In 1977–1982, 33 of 38 nests (87%) successfully produced an average of 1.8 fledglings, whereas in 1983–1989, 42 of 60 nests (70%) produced an average of 1.6 young, with nests closer to human-dominated landscapes producing significantly fewer fledglings than other nests. Human or canine disturbance caused most nest failures.

Diet and feeding behavior. Black Vultures feed opportunistically on human trash, as well as on both small and large wild and domestic animals and carrion. The species prefers fresh versus rotting carrion and also takes vulnerable live prey, including domestic chicks, piglets, and lambs, and feeds on pig feces. In North Carolina, of 117 carcasses fed upon, 11%

were raccoons, opossums, dogs, and wild turkeys, and 89% were livestock, including dead poultry provided by farmers, and pigs and cattle, with poultry being more common in spring and autumn than in winter. Groups of individuals feeding on the carcasses of wild animals were smaller than those feeding on carcasses of domestic livestock. Overall, livestock carcasses and remains appeared to support this non-migrant population.

In urban and other human-dominated landscapes, human-related trash often provides most of the diet. There and elsewhere, Black Vultures depend heavily on the olfactory capacity of Turkey Vultures and often follow them to locate carcasses.

Because Black Vultures are abundant and widespread, and because they routinely feed near humans, an array of somewhat "unusual" items has been reported in their diets. In Louisiana, USA, it is reported to kill and consume striped skunks, even after the latter have discharged their foul-smelling musk. In southern Brazil, Black Vultures clean ticks from the backs of capybaras in the same way oxpeckers remove ticks from ungulates in Africa. It also habitually feeds on vegetable matter including pumpkins, squash, and the fruit of date-palms, with the latter being preferred over carrion in parts of Colombia. In Costa Rica, individuals purposely drag unopened fallen coconuts along roadsides onto well-traveled streets and highways, where trucks and automobiles run over and crush the nuts, exposing the "meat," which the birds then consume.

Only one report estimates the Black Vulture's daily food requirements. Individuals in captivity consumed up to 1.3 lb (600 g) daily.

Social behavior. Much of what we know about the social life of Black Vultures comes from a remarkable 5-year study involving individually marked birds in central North Carolina, USA, in the late 1970s and early 1980s. The study, conducted by the American scientist Patty Parker, followed the species in a 95 mi^2 (250 km^2) mixed agricultural and woodlands site dominated by tobacco, swine, poultry, and cattle farms. Over the course of several years, Parker wing-tagged more than 340 Black Vultures, including 6 radio-tagged birds, among an estimated population of 1200 vultures.

Based on repeated observations of individually marked birds, Parker found that immediate family members maintained close contact year-round and engaged in nearly daily mutual allopreening, communal feeding, and fight intercessions at carcasses.

Typically, young vultures made their first flights 75–80 days after hatching, and they remained in close and almost daily contact with adults at both communal roosts and feeding sites for at least 6–8 months to several years thereafter. Adults typically feed the young for up to 8 months after fledging, and in a few instances, for up to several years. Families that remained together longer tended to have higher breeding success. Closely related individuals were more likely to roost together in the same communal roost on specific evenings. Overall, kin relatedness appeared to be an important force in communal roosts, serving as successful "information centers" for the regional population, and with family coalitions successfully interceding in "food fights" at carcasses that benefited family members. Based on these and other detailed observations, Parker concluded that social and familial organization facilitated "the evolutionary stability of cooperation among communally roosting Black Vultures" at her study site.

Parker's landmark investigation, which remains the only sustained effort of its kind of any species of vulture, underscores the value of careful, long-term, observational studies of marked individuals that will provide meaningful insights into the development and maintenance of social behavior in scavenging birds of prey.

Flight behavior. Black Vultures search for carrion and vulnerable prey in open habitat as well as by following Turkey Vultures that have located carcasses by smell in forested areas.

The wide, low-aspect ratio and slotted wingtips of Black Vultures, which more closely resemble those of Old World *Gyps* vultures than they do other small New World vultures, are typical of species that frequently engage in high-altitude thermal soaring, and appear well suited for that type of flight, as well as for rapid takeoffs from crowded carcass "feeding frenzies" disturbed by approaching mammals.

Movements. Although largely sedentary throughout most of its enormous range, many North American populations retract from the northern latitudinal limits of their range in winter. Short-range movements also occur in response to local and regional shifts in winter weather, and in parts of North America with late-autumn and early winter deer hunting, groups return to feed in areas where "gut piles" are left by deer hunters.

Two American field ornithologists, Eugene Eisenmann and Alexander Skutch, both of whom spent time in the Neotropics in the 1950s and 1960s, independently reported large directional movements of large flocks of Black Vultures in autumn in Panama and Costa Rica, respectively, along the flight lines used by migrating Turkey Vultures that were en route to South America. Although their reports were largely dismissed, because Black Vultures were considered non-migratory at the time, substantial wintertime increases in Black Vulture populations in northern Venezuela support Eisenmann's and Skutch's migration interpretations.

Conservation status. Until close to the end of the 19th century, this human commensal enjoyed a positive relationship with people in cities throughout North and South America where Black Vultures routinely bred in both urban and suburban landscapes. More recently, however, the species has suffered widespread human persecution. In the early to mid-1900s, more than 100,000 Black Vultures were estimated to have been killed in Texas, USA, alone, and today governmental agencies in North, Central, and South America continue to lethally control populations. Black Vultures also are poisoned by consuming poison-laced carcasses targeting mammalian predators of domestic livestock, particularly in central and southern South America.

With global population estimates of considerably more than 10 million individuals, the Black Vulture is considered to be a species of **Least Concern** by BirdLife International.

Turkey Vulture *Cathartes aura*

Spanish: Aura Cabeza Roja
German: Truthhahngeier
French: Urubu à Tête Rouge
aka: Turkey Buzzard

Taxonomy: Family Cathartidae

Size: Length 25–32 in (64–81cm); wingspan 66–80 in (170–200 cm); 3–5 lb (1400–2400 g)
Estimated world population: 13,000,000; increasing
Movements: Partial migrant; some migratory populations long-distance, trans-equatorial and transcontinental
Social ecology: Individual territorial nesting; communal roosting; individual and group food finding

With a distributional range of more than 18 million mi^2 (47 million km^2) and a global population recently estimated at 13 million individuals, the Turkey Vulture is the most widely distributed common vulture in the world, as well as one of the most abundant. One of the least massive of all vultures, the species is an extraordinarily adept soaring machine that feeds on both small and large carcasses and human refuse in many parts of its range, and one that readily adapts to human-modified landscapes. Throughout parts of its extra-tropical range, the Turkey Vulture is the most common bird of prey, and within the tropics shares this attribute of being very common with the Black Vulture. In North America, regional populations fluctuate in response to the availability of carrion ranging from small mammals to beached marine mammals and ungulates. North American populations of the species declined dramatically in the Great Basin of the American West in the 1880s, concurrently with the demise of large herds of American Bison. Subsequently, populations east of the Mississippi River increased concurrently with growing populations of white-tailed deer in the mid-1900s. Many have speculated that increased numbers of roadways throughout North America in the 20th century have also increased populations because of the association with greater amounts of road-killed carrion, including deer.

The species' signature field mark is a prominent dihedral flight profile in which relatively long and slender wings are held above the horizon. This key innovation allows Turkey Vultures to exploit boundary-layer atmospheric turbulence close to the ground, which together with a keen sense of olfaction allows them to both smell and see carrion. Less social than most obligate scavenging birds of prey, and more forest dwelling than many, Turkey Vultures range throughout forested and open habitats across the Americas from south-central Canada in northern North America to southernmost South America. The species thrives on many coastal islands in the Americas, including Vancouver Island, the Bahamas, Cuba, and Jamaica, as well as on Puerto Rico and Hispaniola (where it may have been introduced), and on insular Tierra del Fuego and the Falkland Islands (aka Malvinas).

The Turkey Vulture was described and named *Vultur aura* by the Swedish taxonomist Carolus Linnaeus in 1758, based on a specimen from Veracruz, Mexico. It is grouped with the Lesser Yellow-headed Vulture and Greater Yellow-headed Vulture in the genus *Cathartes*, which has the most species and is the most widely distributed of five genera in the family Cathartidae, a taxon that was once referred to as "pseudo-vultures" based on its distant evolutionary relationship with Old World (Accipitrid) vultures together with anatomical and behavioral characteristics attributable to convergent evolution. The genus *Cathartes* is Greek *kathartes* for "purifier"; the species *aura* is native Mexican for Turkey Vulture.

Subspecies. Turkey Vultures have six recognizable subspecies, which some suggest merit species-level classification given their anatomical differences, such as body mass, wing shape, plumage, the degree of feathering and protuberances on the head, and differences in the coloration of the bare skin of the head and neck. Currently recognized subspecies include *septentrionalis*, a relatively large-bodied, partial, and leap-frog migrant that breeds in eastern North America, populations of which overwinter in Mexico and the southeastern United States; and *meridionalis*, a slightly more massive and darker western North American race, some of which are complete, long-distance, trans-equatorial migrants, whose populations overwinter in Central and South America. Farther south, a decidedly smaller vulture, nominate *aura*, occurs in southern Arizona and possibly peninsular Florida, as well as the West Indies and Central America. Three less-studied subspecies occur in South America. *Ruficollis* occurs in the northern and central parts of the continent; *jota* occurs in parts of southern South America; and *falklandicus* occurs along the coast in the southern cone of South America and its associated islands, including Tierra del Fuego and the Falkland Islands.

Breeding biology. The species does not build a nest, but rather lays typical two-egg clutches in a dark recess, on bare soil, stones, and gravel, in rocky crevasses, fallen timber and snags, unused mammal burrows, the bases of brush piles and thick ground vegetation, and the floors of abandoned and near-abandoned buildings. In one instance a pair nested in the open trunk of an abandoned car.

Detailed studies of the nesting biology of Turkey Vultures are uncommon. One, in Saskatchewan, Canada, in the early 2000s, merits note. Instances of vulture nests in the region date from the 1980s, when several nests were first reported in the attics of abandoned farmhouses. One hundred and twenty-six nesting attempts in 74 deserted buildings in 2003–2006 averaged 1.7 fledglings, which is breeding success similar to those reported in Pennsylvania, Maryland, and Wisconsin, USA. Nesting failures included accidental deaths, predation, and human disturbance.

Despite increased numbers during the study, no building held more than two nests, suggesting territoriality.

Diet and feeding behavior. The species feeds opportunistically on organic human trash, mammals, reptiles, and both adult and nestling birds, as well as on wild and domestic small to large carrion found near and far from active nests. Evidence suggests Turkey Vultures prefer relatively fresh carcasses, but consume putrefied matter when other options are not available. Occasionally, the species also takes small and large living prey, including fish in desiccating puddles and fallen sick or injured sheep. It also induces regurgitation in recently fed nestlings of large birds, including herons. A slow and meticulous feeder, Turkey Vultures are capable of stripping organic matter from carcasses, including snakes and other small vertebrates, without disarticulating their carcasses.

Three factors expand the composition of the species' diet beyond that of other obligate scavenging birds of prey. First, they have an enormous geographic range that includes both the Neotropics and northern and southern New World temperate zones. Second, individuals often search both freshwater and marine drift lines in search of carcasses, a behavior that increases their exposure to water-dwelling carrion that ranges in size from small marine invertebrates to enormous marine mammals including whales. Third, the species' olfactory ability allows individuals to seek out and find small and large carcasses in both forest-canopied and open habitats. Overall, domestic livestock, and the invertebrate fauna associated with it, dominate the species' diet in agricultural areas, whereas native species tend to dominate the diet in more natural areas. Where Turkey Vultures occur together with Black Vultures, the former tend to feed on smaller carrion. Diets based on regurgitation pellets collected during two winters at a mixed species roost of approximately 70% Turkey and 30% Black Vultures in Pennsylvania, USA, indicated that larger carcasses, including both livestock and deer, were more common than small mammals and poultry during times of snow cover, most likely because larger carrion were more conspicuous then.

Based on measurements from captive birds held in cages in their thermal-neutral zone and fed chicks or hamsters, individual Turkey Vultures consume 6% of their body mass, or approximately 200 kcal of metabolizable energy, daily. Captive individuals deprived of food and water lost 18% of body mass over 8 days with no apparent "ill affect," but one individual died that was not fed for 20 days and lost 20% of its body mass.

Social behavior. Although less social than many species, Turkey Vultures typically roost communally, sometimes by the hundreds, both with and without other species. Turkey Vultures roost in trees and on human constructions, including buildings and on electric-line poles and stanchions, as well as on the ground near large garbage dumps. As is true of other species of vultures, soaring Turkey Vultures descending onto carcasses often attract the attention of other individuals, including conspecifics and Black Vultures. Adults usually dominate juveniles at such times, and Black Vultures typically behaviorally dominate Turkey Vultures, particularly when the former occur in large numbers. In northern South America, where larger migratory races of overwintering North American Turkey Vultures seasonally co-occur with resident Black Vultures, the former dominate the latter, presumably because the migrants are larger than local residents, which results in a reversal in relative body masses in Turkey Vultures versus Black Vultures. In the Falkland Islands, Turkey Vultures dominate Striated Caracaras at carcasses in austral summer, but they in turn are dominated by caracaras in winter, particularly when feeding together with groups of food-stressed juvenile caracaras.

Flight behavior. Turkey Vultures search for carrion in both high- and low-altitude soaring and gliding flight, accentuated with characteristic side-to-side rocking flight that is far more apparent when individuals fly low over vegetation. On migration, the species typically engages in soaring interspersed with gliding flight on partially flexed wings in sometimes multi-thousand bird flocks. Turkey Vultures depart mixed-species roosts earlier in the morning than do Black Vultures, presumably because lighter wing loading allows them to soar in weak, early morning updrafts.

Movements. Turkey Vultures are the most migratory of all vultures, with as many as 3 million individuals moving from breeding areas in North America to wintering areas south of the equator and traveling in enormous flocks (Box 2.1).

Box 2.1 Tagging and Migration of Turkey Vultures

Like other New World vultures, Turkey Vultures routinely defecate on their legs, which is a behavior that has compromised banding efforts. Turkey Vultures frequently stand on the rotting carcasses they feed upon, which serve as ideal breeding grounds for flesh-eating microbes. Although the antiseptic nature of vulture defecation remains to be tested, its highly acidic nature suggests that it may help a vulture's feet and legs stay free of infections that could otherwise develop in cuts and scratches. Unfortunately, such *urohidrosis* has the potential to hobble a vulture if uric acid builds and "cements" to a leg band (aka ring). For this reason, both Canada and the United States have prohibited banding for decades. Many believe vultures engage in urohidrosis—the literal translation from ancient Greek is to "sweat urine"—to evaporatively cool themselves in the heat of summer. But given that at least Turkey Vultures sometimes behave in this manner in winter suggests an additional, antibacterial, function.

Fortunately, patagial wing tagging, satellite tracking, and the establishment of several million-bird hawk watches in Mexico and Central America in the 1990s have rekindled an interest in Turkey Vulture migration, and scientists are beginning to understand the geographic scope, magnitude, and nature of this species' migratory journeys. The emerging picture suggests that the Turkey Vultures' largely free-of-feeding, long-distance trek between North and South America ranks among the most extraordinary of all raptor migrations.

Like most vultures, Turkey Vultures are both social and, at times, gregarious. Most, for example, roost communally year-round, often in the company of Black Vultures. Individuals also group when migrating. In autumn on outbound migration, flocks of long-distance migrants build quickly from several individuals to several dozen individuals, and soon thereafter swell to multi-hundred and multi-thousand bird aggregations along major corridors and flyways. The same occurs on return migration in spring. Short-distance migrants coalesce into smaller groups.

In both eastern and western North America, long-distance migratory Turkey Vultures begin to move in late August. The birds spend time feeding early in migration, and the bulk of the outbound flight occurs much later than this, particularly in eastern North America, where overall distances traveled are much shorter than in the American West. At the Hawk Mountain Sanctuary watch site, in eastern North America, more than half of all migrating Turkey Vultures are counted after 20 October, and in many years the largest flights occur in November.

The most significant North American flight east of the Mississippi River occurs in southern Ontario and southeastern Michigan, where 10,000 to

20,000 migrating vultures concentrate along the northern shorelines of Lake Erie while being diverted southwest. Numbers in the region peak in mid-October, with daily counts exceeding thousands of birds in most years.

The real action, however, takes place in the American West, where trans-equatorial *meridionalis* subspecies makes its appearance. Hints of a flight begin in mid-September in coastal British Columbia and in Washington State, where more than a thousand Canadian breeders travel across the 20-km Strait of Juan de Fuca from the southern tip of Vancouver Island in British Columbia to make landfall in northwestern Washington. Several weeks later, and approximately 1600 km farther south, movements have built to 30,000 individuals at the southern tip of the Sierra Nevada in southern California. This, and other vulture movements from farther south, converge as a growing queue of individuals flows southeast along the slopes of the Sierra Madre Occidental of northern Mexico. Somewhere south of Monterrey, Mexico, the flight joins similar growing numbers of outbound Broad-winged Hawks and Swainson's Hawks. Shortly thereafter, the three species converge north of the world's narrowest and most significant migration bottleneck near Cardel, in Veracruz, Mexico, where 1.5 million to more than 2 million southbound Turkey Vultures are counted in October.

Most individuals continue south across the Isthmus of Tehuantepec, before turning westward and following the foothills along the Pacific slopes of southern Mexico, Guatemala, El Salvador, Honduras, and Nicaragua. In northern Costa Rica, the birds cross the Continental Divide for a third time and soar above the Caribbean slopes of that country's Talamanca Mountains before entering Panama and, after that, northeastern Colombia.

The South American wintering grounds of western North America's 2–3 million Turkey Vultures remain something of a mystery. It almost appears as if the birds simply "vanish" upon reaching South America. Part of the mystery is that South America's resident Turkey Vultures make it difficult to determine where the migrants settle each autumn; another part is because Turkey Vultures are not banded in North America, and, hence, no band recoveries can be observed on the wintering grounds.

The lack of banding recoveries notwithstanding, anecdotal reports, patagial wing tagging, and satellite tracking indicate that many North American Turkey Vultures overwinter in Panama, Colombia, and Venezuela, rather than continuing farther south. Some of these overwinter in the expansive freshwater wetlands or Llanos of central Venezuela, where identifiable migrants of the *meridionalis* subspecies race are known to dominate members of the smaller resident *ruficollis* subspecies during aggressive encounters, and to eventually replace them at the latter's preferred feeding and roosting areas. That the North American subspecies arrives at the onset of the dry season—a time when reptilian and mammalian mortality increase substantially—enhances carcass availability for both races. Even so, resident individuals forego breeding until after the migrants leave in spring.

Although the geography of Turkey Vulture migration is reasonably well understood, energy management during their long-distance flights is not. One thing is certain, no significant feeding occurs en route, particularly south of the United States. With 80% of the outbound flight of more than 1 million Turkey Vultures passing southern Costa Rica in fewer than 2 weeks each autumn (i.e., at a rate approaching 100,000 individuals daily), it seems highly unlikely that individual birds can find and consume food while traveling in such close quarters. To offset this lack of feeding, indications suggest that migrants expend remarkably little energy while traveling long distances.

My own observations, in both Pennsylvania and in Costa Rica, suggest that once individuals have been airborne for 5–10 minutes each morning (an initial period of flight in which birds seek and find their first thermal or updraft), migrants travel solely by slope and thermal soaring for the remainder of the day, flapping rarely, until a few minutes prior to the end of the day's flight, when they descend into roosts.

Smithsonian Tropical Research Institution researcher Neal Smith, who has studied the species' migrations in Panama, along with those of migrating Broad-winged and Swainson's Hawks, rated the vulture's soaring behavior as vastly superior to that of the two buteos. Nevertheless, migrants arrive in Venezuela during late autumn in poor condition and need to rebuild their body mass during their stay before returning north the following spring. A combination of accumulated body fat, excellent flight skills, abundant solar energy in the form of near-continuous thermals, and a likely drop in core body temperature at night are what enable western North American populations of Turkey Vultures to successfully complete an approximately 8700-mi (14,000-km) round-trip journey each year.

As is true of many vultures, the home ranges of Turkey Vultures vary enormously, both geographically and temporally, with larger ranges occurring outside of the breeding season. The ranges of aerial and satellite-tracked non-breeding migratory Turkey Vultures captured in communal roosts in South Carolina shifted by more than an order of magnitude, from less than 40 mi^2 (100 km^2) to more than 480 mi^2 (1200 km^2), and those captured on nests in Saskatchewan, Canada, at the northern limit of its range, ranged from less than 200 mi^2 (518 km^2) to more than 740 mi^2 (1900 km^2). Home ranges of Canadian migrants overwintering in Venezuela varied by four orders of magnitude from 20 to 30,000 mi^2 (54 to 76,700 km^2). Monthly home ranges of 9 non-migratory Turkey Vultures monitored for at least 1 year in South Carolina, USA, averaged 24 ± 1.7 mi^2 (62 ± 4.3 km^2), or approximately twice that of 9 Black Vultures monitored at the same site, with ranges varying little from month to month except for a substantial drop in May, when parents were caring for recently hatched young.

Conservation status. With a global population recently estimated at 13 million birds, the species is one of the most successful of all obligate scavenging raptors. In the 1800s, the species was welcome in human-dominated landscapes in colonial North America where it was prized for controlling carrion and organic waste. It was culled in parts of North America in large numbers in the early and mid-1900s by farmers and rural residents in the southeastern United States, in part because of the mistaken fear of it spreading livestock and human diseases. As of 2019, individuals were still being removed in the southeastern United States as part of governmental actions directed at the more "problematic" Black Vulture populations. The species is killed in Central and South America by poison-laced carcasses set for mammalian predators of livestock. Given recent catastrophic declines in previously widespread and abundant populations of Old World Vultures, additional population monitoring is warranted.

The Turkey Vulture is considered to be a species of **Least Concern** by BirdLife International.

Lesser Yellow-headed Vulture *Cathartes burrovianus*

Spanish: Aura Cabeza Amarilla
German: Kleiner Gelbkopfgeier
French: Urubu à Tête Jaune
aka: Savanna Vulture

Taxonomy: Family Cathartidae

Size: Length 23–26 in (58–66 cm); wingspan 63 in (160 cm); 2 lb (950 g)

Estimated world population: c. 100,000; possibly decreasing but likely stable, no indication of population increases anywhere in its range

Movements: Largely sedentary; sometimes nomadic

Social ecology: Individual nesting; communal roosting, sometimes with other species; largely individual food finding

One of the least studied of all vultures, Lesser Yellow-headed Vultures have a 4.3 million mi^2 to 7.4 million mi^2 (11–19 million km^2) tropical range in 21 countries. A slender version of the conspecific Turkey Vulture, Lesser Yellow-headed Vultures have blackish plumage overall, highlighted with a greenish or bluish gloss. The species' narrow and somewhat pointed wings are held in a decided dihedral and feature white-shafted primaries that superficially resemble those of Black Vultures. With at least some representatives weighing less than a kilogram, the species ranks among the least massive of all vultures. Its common name, which is applicable to living birds but fades quickly in death, refers to its distinctive yellow-to-orangish, bare-skinned head with scattered blue markings, the nuanced details of which are described below by the American museum curator Alexander Wetmore in 1950, based on a freshly killed Argentine specimen:

> The bill was cream buff, shading to vinaceous buff on a broad area that extended onto the forehead. Behind the nostrils; side of the head in general, including eyelids, deep chrome; center of the crown Tyrian blue, bordered on either side by a broad band of stone green; skin of throat posteriorly deep chrome, becoming paler forward, to shade into olive buff toward base of bill; space between mandibular rami spotted with dark Tyrian blue; a dull spot of slate blue beneath the nostrils on either side; iris carmine; tarsus cartridge buff, shading to neutral gray on the toes, where the interscutal spaces have a scruffy whitish appearance.

The species has numerous caruncles (i.e., small, fleshy lumps or swellings in the skin), most notably on the hind crown and bare parts of the neck, similar to those on both Turkey Vultures and Greater Yellow-headed Vultures. Although the functional significance of its distinctive head coloration has yet to be determined, some attribute it to individual recognition indicative of age or dominance status, or both. Unless well studied in the field, the species' distinctive head coloration and markings are often overlooked, and its occurrence underreported because of its superficial resemblance to the more widespread, and typically more common, Turkey Vulture.

Much of what has been reported about the Lesser Yellow-headed Vultures comes from brief species accounts in summaries of regional avifauna and from anecdotes of unusual occurrences.

The species was named in 1845 by collector Marmaduke Burrough and thereafter described by the American museum ornithologist John Cassin, based on an individual collected near Veracruz, Mexico. The genus *Cathartes* is the Greek *kathartes* for "purifier"; the species *burrovianus* is after collector Burrough.

Subspecies. Two subspecies are recognized: the smaller nominate *burrovianus* occurs in Central America and northern South America, and the larger *urubitinga* occurs in South America from southeastern Colombia and Guyana south.

Breeding biology. The Lesser Yellow-headed Vulture nests in the cavities of large snags and on the ground in the cavities of fallen brush, trees, and dense tussock grass. Breeding seasons vary regionally. The most detailed published study of breeding occurred in 1995–2004 in the 8243 ac (3336 ha) El Bagual Ecological Reserve in the seasonally flooded humid Chaco of the province Formosa in northeastern Argentina. During this study, activities at 13 ground nests of this locally "common" resident were monitored. Nesting occurred in early September through early February. Incubation of two-egg clutches lasted approximately 40 days, and the young remained at or near the nest for 70–75 days. At least one clutch was disturbed by Southern Crested Caracaras after a nest visit by a worker. The only other predator encountered was the Argentine black-and-white tegu, a large lizard.

Diet and feeding behavior. Lesser Yellow-headed Vultures, like other *Cathartes*, search for food both by sight and by smell, and do so both alone and in small groups. Lesser Yellow-headed Vultures feed principally on small prey, including fish, amphibians, reptiles, and small birds and mammals. The species feeds sporadically with other vultures on larger carcasses, including beached marine mammals, and on fish stranded at the perimeters of seasonally ephemeral wetlands. Some suggest that its extremely low-altitude flight is an adaptation for surprising and taking live prey, and there are reports of the species catching and consuming living snakes in Mexico and Brazil, including a venomous Brazilian lancehead, which it subdued by repeatedly pecking its head.

The extent to which it competes with larger vultures remains largely unstudied. Lesser Yellow-headed Vultures have measureably wider bills and larger gapes than do Turkey Vultures, which may allow them to consume carrion more quickly. The species is quite tame, which may inrease its access to carrion in human-dominated landscapes.

Social behavior. Lesser Yellow-headed Vultures often roost together with Turkey Vultures and Black Vultures. It breeds in loose colonies in northeastern Argentina.

Flight behavior. Published descriptions, together with my own observations in Costa Rica and Uruguay, indicate that the species flies closer to the vegetation—often to within a meter or less—than do Turkey Vultures. Some attribute this phenomenon to its lighter wing loading, which may explain why it usually arrives at carcasses before Turkey Vultures.

Movements. Individuals that breed in Central America are said to move south into northern South America during the non-breeding dry season. In December–March, the species occurs in greater numbers in northern Colombia and outnumbers Turkey Vultures in the savannas of Venezuela. Belizian breeders may travel west into Mexico in February–July, and there are reports of hundreds flying southwest into Rio Grande do Sul, Brazil, in April.

Conservation status. Populations in Uruguay, where Black Vultures are purposely poisoned by sheep ranchers, may be limited. With global population estimated at 100,000 to <5,000,000 individuals, the Lesser Yellow-headed Vulture is listed as a species of **Least Concern** by BirdLife International.

Greater Yellow-headed Vulture *Cathartes melambrotus*

Spanish: Aura alas Anchas, Aura Silvatica
German: Großer Gelbkopfgeier
French: Grand Urubu
aka: Forest Vulture

Taxonomy: Family Cathartidae

Size: Length 25–30 in (64–75 cm); wingspan 66 in (170 cm); 3.6 lb (1.65 kg)

Estimated world population: 10,000–100,000; possibly decreasing but most likely stable, no indication of population increases anywhere in its range

Movements: Largely sedentary; some wandering

Social ecology: Individual nesting; communal roosting, sometimes with *C. aura*; largely individual food finding

Less well studied than the congeneric Lesser Yellow-headed Vulture, this bulkier and entirely neotropical species occurs in nine Amazon-basin countries and has a distributional range estimated at 2.7 million mi^2 (7 million km^2).

The species was first reported to science based on an adult specimen collected by Pinney Schiffer in British Guiana in 1950. A more robust and broader-winged version of the Turkey Vulture, the Greater Yellow-headed Vulture has deep black plumage with an iridescent greenish sheen and dull brownish underwings and tail. The wings, which are held in a decided dihedral, feature white-shafted primaries above. Its common name, which is valid for the living bird but not for museum specimens, refers to its distinctive yellow-to-orangish, bare-skinned head whose yellow coloration extends to the malar region, unlike the Lesser Yellow-headed Vulture. The functional significance of its distinctive head coloration, which fades in museum specimens and has yet to be studied in detail, may be linked to individual recognition.

As in the Lesser Yellow-headed Vulture, much of what is known of the species behavior and ecology comes from brief species accounts in regional avifauna publications and from anecdotal reports of unusual distribution or behavior.

The genus *Cathartes* is the Greek *kathartes* for "purifier"; the species *melambrotus* is Greek for "blackish."

Subspecies. No subspecies are recognized.

Breeding biology. Greater Yellow-headed Vultures nest in the cavities of large trees, at least sometimes.

Diet and feeding behavior. As in other *Cathartes*, the species uses both smell and sight to locate carcasses on the forest floor, including those of primates, sloths, and opposums. Greater Yellow-headed Vultures are dominated by the more massive and more stoutly billed King Vulture, but not necssarily by Turkey Vultures. Greater Yellow-headed Vultures often follow King Vultures to carcasses, presumably to take advantage of the latter's abilty to open large carcasses.

Flight behavior. Most descriptions indicate that the species flies routinely low over forest canopies.

Movements. Individuals are believed to be sedentary, although wandering has been reported in lowland Bolivia.

Conservation status. Population estimates range from 100,000 to <5,000,000. The Greater Yellow-headed Vulture is listed as a species of **Least Concern** by BirdLife International.

King Vulture *Sarcoramphus papa*

Spanish: Zopilote Rey
German: Königsgeier
French: Sarcoramphie roi

Taxonomy: Family Cathartidae

Size: Length 28–32 in (71–81 cm); wingspan 70–79 in (180–200 cm); 6.5–8.4 lb (3.0–3.8 kg)
Estimated world population: <10,000; possibly decreasing
Movements: Sedentary; sometimes nomadic
Social ecology: Individual nesting: communal roosting; individual and group food finding

A denizen of extensive lowland forest, the King Vulture ranges over slightly more than 7.8 million mi^2 (20 million km^2) of forest and savanna from central Mexico, south through the Amazon Basin. The species is intermediate in size between the smaller *Cathartes* and *Coragyps* vultures and the larger Andean and California condors. King Vultures are most conspicuous when soaring over open areas near forest edges and when feeding on the ground. Although usually not the first to arrive at carcasses, the species' massive bill is said to allow it to more easily open larger carcasses than the decidedly smaller bills of smaller species.

The breeding and movement behaviors of King Vultures are not well understood.

With a brilliantly detailed and garishly festooned featherless head and neck, and a striking black and creamy white plumage, adult King Vultures are unmistakable in the field. Recently fledged young are sooty black throughout. The transition from immature to full adult plumage takes 4 to 5 years. Although the degree to which individuals use variation in the head and neck appearance to identify each other has not been studied, human keepers of captive individuals use differences in caruncle and wattle size to identify individuals, and presumably, so do King Vultures. The extent to which their ornate appearance also signals behavioral intent or physiological condition remains unanswered.

Included in the condor clade of New World vultures that split from the *Coragyps-Cathartes* clade approximately 14 million years ago, and from the California Condor more than 3 million years ago, fossil individuals appear in late Pleistocene sediments in central Argentina. The King Vulture was described as *Vultur papa* by Carolus Linnaeus in 1758, and renamed *Sarcoramphus papa* in 1806. The genus name *Sarcoramphus* is Greek for "fleshy bill"; the species name *papa* is Latin for "bishop."

Subspecies. No subspecies are recognized.

Breeding biology. King Vultures are largely ground and cliff-nesters that sometimes nest in snags. Incubation of the typical single-egg clutch takes 50–58 days. Fledging occurs at approximately 130 days. Parental presence at nests is thought to be greatest early in the day. Factors affecting nesting success are unstudied.

Diet and feeding behavior. The species does not find carrion by smell, unlike the *Cathartes* vultures, and they typically arrive at carcasses after *Cathartes* and *Coragyps* vultures, possibly by following the former to carcasses they have located olfactorily. Field evidence suggests King Vultures search for carcasses in pairs and family groups. They dominate *Cathartes* vultures at carrion but are intimidated by groups of Black Vultures. Its heavy bill is capable of opening the carcasses of large animals better than the smaller and lighter bills of smaller vultures, and the latter take advantage of this difference when feeding together. Several researchers have suggested that smaller vultures find and then purposely lead King Vultures to large carcasses they have found; however, field observations have yet to confirm this relationship. Free-ranging Turkey Vultures roosting at

Venezuelan zoos routinely perch by the dozens above display enclosures of captive King Vultures, but not above the enclosures of other captive avian scavengers, including Andean Condors, which supports this symbiosis.

One of the more intriguing series of food-finding observations in the species are those of the French New World vulture specialist Marsha Schlee who studied their movements in southern Venezuela, where as many as six individuals fed upon the carcasses of both livestock and wild animals across several years. Schlee followed the birds as they circle-soared over hunting jaguars, and possibly a puma, that were stalking white-tailed deer and other potential prey, the carcasses of which were then scavenged by the vultures. This finding suggests that King Vultures sometimes track mammalian predators.

Like other New World species, King Vultures sometimes take vulnerable small animals. It also feeds on the fruits of several species of native palms.

Social behavior. King Vultures roost in family groups of as many as eight or more individuals. No evidence suggests that it roosts with other vultures. The extent to which immature individuals remain with their parents is unstudied.

Flight behavior. The soaring behavior and extent of areas used by King Vultures for thermal soaring over the western Orinoco Basin in Venezuela were tracked from a small, fixed-winged aircraft across several years in the 1980s. Of the birds studied, 80% traveled alone, and the remainder were in pairs or small groups of adults and young. Flight altitudes over forests average between 850 and 950 ft (260 and 290 m) in the wet and dry season, respectively, with most observations beginning in midmorning.

Movements. The species has not been wing tagged or satellite tracked, and little is known of its movement behavior.

Conservation status. Several factors suggest that the species' conservation status needs to be updated. First, because it is a forest habitat specialist. Second, because large-scale studies of factors affecting its distribution and abundance have yet to be conducted. Third, because its seasonal movements and ranging behavior have yet to be investigated. And fourth, because of suggested substantial population declines in many parts of its range. In spite of suggestions of a declining global population of <10,000 mature individuals, and a relatively small distributional range of 7.8 million mi^2 (<20 million km^2), the species is currently listed as **Least Concern** by BirdLife International.

California Condor *Gymnogyps californianus*

Spanish: Cóndor Californiano
German: Kalifornischer Kondor
French: Condor de Californie
aka: California Vulture (obsolete)

Taxonomy: Family Cathartidae

Size: Length 46–53 in (117–134 cm); wingspan c. 106 in (270 cm); 18–31 lb (8–14 kg)
Estimated world population: <1000; increasing
Movements: Sedentary; sometimes nomadic
Social ecology: Individual nesting; communal roosting; individual and group food finding

The California Condor is the most massive and largest vulture in North America, as well as one of the most thoroughly studied of all obligate scavenging raptors. As a result of the ups and downs of its conservation history and the many scientific studies associated with it, we know more about the limiting factors associated with its decline than of any other vulture. As such, its basic biology, together with the often complicated and sometimes tortuous history of its conservation, illustrate considerable insight into how best to surmount the challenges of conservation success.

Considered by some to be a Pleistocene relic associated with North America's extinct megafauna, the species that once ranged across much of western North America, east into Peninsular Florida and the Great Lakes, now has a global distribution limited to fewer than an estimated 70,000 mi^2 (180,000 km^2) in southern California, northern Arizona, and southern Utah, USA, and northern Baja California, Mexico. With an increasing population of fewer than 600 individuals in late 2019, the California Condor is, by far, the least common of vultures. Currently their population is largely non-migrant and depends heavily on feeding stations established to provide them with clean, lead-free food. The California Condor once fed on North American megafauna, and more recently on domestic and wild ungulates inland, and marine mammals and anadromous fish coastally.

The remains of condor feathers and other body parts have been found in the burial mounds of Native Americans, many of whom used condor feathers in their ceremonial clothing. The Lewis and Clark Expedition of 1804–1806, which explored the then-recently acquired Pacific Coast of the United States, reported many condors. They shot at least five individuals near the mouth of the Columbia River in Oregon and immediately recognized the species as the largest vulture in North America. In the latter 19th and early 20th centuries, prospectors panning for gold may have significantly reduced populations when condors were shot for their quills for use in holding gold dust.

In the face of catastrophic declines in the 1980s that were generally, but not universally, attributed to lead toxicity, the remaining remnant wild population of California Condors was brought into captivity and a rigorous captive-breeding scheme was launched to help save the species from extinction (Box 2.2).

Box 2.2 California Condor Conservation Timeline

During the Pleistocene period, several species of the genus *Gymnogyps* occurred along parts of both the East and West Coasts of North America as well as in the southwestern United States before its range contracted to the West Coast.

1792	First California specimen collected in Monterey, California, USA
1804–1806	Sighted and collected by Lewis and Clark in the Pacific Northwest, USA
1850	Largely extirpated from Washington and Oregon, USA
1905	Killing or collecting of condors or their eggs made illegal by the legislature in California
1906	First detailed study of a nesting pair
1939	Systematic studies began by University of California graduate student Carl Koford
1950s	Condors confined to southern California
1950s	Unofficially, the California Fish and Game Commission estimates the wild population at more than 50 individuals

1970s	Condor Recovery Team established by the United States government
1978	American Ornithologists' Union–National Audubon Society panel recommends capturing all birds
1982	22 condors estimated to exist in wild
1987	Last wild bird brought into captivity
1988	Successful reproduction in captivity
1992	First release of captive-bred birds
1994	All released birds captured due to behavioral problems
1995	Releases begin in southern California
1996	Releases begin in Arizona
1997	Releases begin in Big Sur, California
2001	First attempted breeding in the wild
2002	Releases begin in Baja California
2004	First chick fledged in the wild
2008	Lead ammunition banned in condor range in California
2020	More than 500 condors, with 400-plus in the wild

Adults are blackish overall, with streaked gray ruffs and white-fringed greater wing coverts. Their featherless heads are wrinkled, red-orange to yellow, with a black bristly forecrown, pinkish to reddish neck, and their wing linings are whitish below. Juveniles are similar to adults except that their heads and bodies are largely dusky and covered with short, fuzzy down on the neck, with black mottled wing linings below.

The species was named *Vultur californianus* in 1797 by the British naturalist George Shaw, based on a specimen collected by the British botanist Archibald Menzies in coastal California during the British Vancouver Expedition of 1791–1795, and was renamed *Gymnogyps californianus* shortly thereafter. The genus name *Gymnogyps* is Greek for "naked" and "vulture"; and the species name *californianus* is Latin for the site of the type specimen. The vernacular name "condor" is from the Quechua (Inca) word, "cuntur."

Subspecies. The relictual nature of the late 20th-century population available for study precludes a detailed genetic analysis. In the early 1990s, DNA fingerprints from 32 individuals of the remnant wild population that had been brought into captivity suggested low genetic diversity within three sibling-adult subgroups. They exhibited substantially lower among-group relatedness than within-group relatedness, a factor that conservationists used in assigning breeding pairs in captivity.

Breeding biology. Detailed observations of the breeding biology of California Condors date from March of 1939 when the American naturalist Carl Koford began graduate student studies at the University of California, Berkeley. Koford visited 15 condor nest sites, all but one of which were in cool grasslands or woodlands in the Upper Sonoran Life Zone, 500–1500 yd (c. 500–1500 m) above sea level (asl) in southern California. All of Koford's sites were in cave-like, crevasse, or boulder-pile cavities, and all were above ground level, presumably so that birds could descend in gliding flight when exiting them. Openings were 0.5–0.8 yd (c. 0.5–0.8 m) wide and up to 5 yd (4.5 m) from the cave entrance.

Prior to the 1980s only one nesting effort had been observed meticulously for more than 100 days. That changed in the 1980s, during the era of significant population decline in California. Continual observations were made at six nests, and findings showed that pairs were loyal to both their mate and territory with no instances of divorce, and with replacement mates occurring only when a pair member had died. In general, individuals were not aggressive when approached by humans, but appeared curious or fearful. Nesting territories tended to be held year-round. Courtship behavior, including "coordinated pair flights, mutual grooming, and sexual display," usually occurred from late autumn through early spring, with a tendency for displays to precede copulations, which occurred shortly before egg laying. The female typically "filled her crop" with water a day or two before egg laying, something that also has been observed in captivity. At least one female laid an egg while standing, with the egg suffering no apparent damage. Both pair members incubated, with individual shifts lasting as long as 9 or 10 days.

Hatching was protracted, with pipping taking close to 3 days, and with adults sometimes gently breaking off bits of shell during the process. Adults attended nests continuously during the first month, with shifts averaging 3.9 days. At first, parents fed nestlings almost daily, with each parent feeding the chick about five times weekly. So-called begging wing flaps by the nestling appeared to help parents locate their young. Nestlings fledged 5 to 6 months after hatching, usually across mid-September through early November, making clumsy initial flights that ranged from 20–300 yd (c. 20–300 m).

Nesting typically occurred in areas with relatively low populations of Golden Eagles. Northern Ravens, which were decidedly more common than Golden Eagles, were often chased from nesting sites. Ravens were seen entering nest caves, and at least one destroyed an unattended egg.

During the early 1980s, most studied adults were paired and most attempted breeding. Overall, nesting success did not appear to affect the population's ongoing decline.

Diet and feeding behavior. Based on recent isotopic analyses, populations are believed to have retracted from many inland areas during the late Pleistocene and early Holocene as populations of North American megafauna declined. Coastal populations survived by feeding on aquatic and marine resources, including the carcasses of salmon and marine mammals. In the first decade of the 1800s, Lewis and Clark reported condors feeding near the mouth of the Columbia River on the carcasses of fish and whales that had been "thrown up by the waves on the sea coast." Other early observers reported condors feeding on salmon during spawning runs farther inland. Shortly thereafter, the American naturalist A. S. Taylor noted hundreds of individuals feeding on the carcasses of northern sea lions slaughtered for their oil during the era's seal rush. In the 1850s, the American naturalist and ornithologist Spencer F. Baird reported that correspondents saw them feeding on seals and whales along the coast. Additional accounts indicate a dietary shift toward cattle and deer late in the 1800s. Today, most condors feed at condor feeding stations established by conservationists to reduce exposure to lead shot. That marine subsides likely facilitated the species' survival through late-Pleistocene population declines reflects condors being in the right place (i.e., coastally) during the Pleistocene megafaunal extinction event.

Social behavior. Although assessing the social nature of California Condors is compromised by current small and disarticulated reintroduced populations, historical observations suggest at least episodic communal roosting and group feeding in coastal populations. Reports from the early 20th century describe dozens of condors being flushed from a fresh deer carcass, and as many as 30 individuals landing to feed on a single deer or steer carcass. As recently as the 1940s, 22 birds were reported feeding on a single calf carcass. In the early 21st century, dozens of birds routinely visit individual feeding stations, leaving little doubt that the species often fed communally on large carrion. Although the extent to which the species uses social networks to locate and take advantage of fresh carcasses awaits analysis of movements of individually satellite-tracked birds. Historical observations suggest a reasonable degree of social feeding.

As the number of parentally reared wild birds entering growing populations continues to increase, the extent to which both intra- and inter-generational learning influences social behavior is likely to become evident. Limited inter-generational learning has caused, and continues to cause, significant problems for many juveniles, and the ongoing reintroduction effort continues to provide opportunities to study the extent to which social behavior in California Condors affects population success.

Movements. The extent to which historical populations were migratory is open to debate. One early 19th-century observer, the Scottish botanist David Douglas, reported "great numbers" along the Columbia River seasonally, with lower numbers in winter. The American naturalist John Kirk Townsend also recorded them feeding along the Columbia on dead salmon, but again only in spring and summer, with individuals "retiring, probably to the mountains, about the end of August," and the Canadian naturalist John Fannin characterized them as "very rare summer residents in coastal British Columbia." The American naturalist William Gambel noted that they were more abundant in California in winter, while linking their increased numbers in that season to coincidental declines in the more northerly Columbia River population. Given that nutritionally significant resources of condors may have shifted seasonally over the past 150 years, and that like Andean Condors, California Condors likely depend on strong updrafts for both short- and long-distance travel, suggest that migratory movements are probably limited to relatively short distances.

Satellite tracking indicates that California Condors do have large home ranges, that pairs commonly travel together when seeking food, and that breeding-season movements typically are limited to within 85 mi (137 km) of nest sites. A recent study that tracked 74 individuals documented annualized monthly home ranges averaging 220 mi^2 (570 km^2) in adults and 160 mi^2 (410 km^2) in immatures, with cross-country speeds of up to 55 mph (90 km/hr), and daily linear ranging movements sometimes exceeding 600 mi (965 km). Late summer and early autumn home ranges are most likely larger because of stronger thermals during that time frame, and are smaller during the breeding season because of nest-associated duties.

Conservation status. Although considered tolerably common by naturalists well into the late 1800s, the species was being poisoned incidentally by ranchers targeting predatory mammals. As early as 1890, the American Army surgeon and naturalist James Cooper branded the species as "doomed," stating that it already was "in the process of extinction." Shortly thereafter, others began to take notice. The National Audubon Society commissioned the first thorough study of condor behavior and ecology in 1939, and in the 1980s the remaining 22 wild birds were taken into captivity and a monumentally heroic and expensive governmental and private captive-breeding and reintroduction program initiated. By 2019, more than 330 individuals existed in the wild in parts of California, Arizona, Utah, and northern Baja California, Mexico. Currently, the principal limiting factor threatening the species is widespread lead poisoning due to lead ingested in hunter-shot game.

Given the species' drastic decline in the latter 20th century and its current dependence on intensive conservation management, the California Condor is listed as **Critically Endangered** by BirdLife International.

Andean Condor *Vultur gryphus*

Spanish: Cóndor Andino
German: Andenkondor
French: Condor de Andes

Taxonomy: Family Cathartidae

Size: Length 40–51 in (100–130 cm); wingspan ≥126 in (≥320 cm); 18–33 lb (8–15 kg)
Estimated world population: <15,000; decreasing
Movements: Sedentary; occasionally nomadic and migratory
Social ecology: Individual nesting; communal roosting; individual and group food finding

The Andean Condor is the world's largest obligate scavenging raptor, as well as one of the most massive of all flying birds. The national symbol of Argentina, Bolivia, Chile, Colombia, Ecuador, and Peru, the species ranges altitudinally to at least 16,400 ft (5000 m) across 3.3 million mi^2 (8.5 million km^2) of its range along the mountainous Andean spine of western South America from Venezuela and Colombia, south into Chile and Argentina. A sedentary species with large territorial ranges, condors currently depend heavily on the carcasses of domestic livestock.

Charles Darwin considered the Andean Condor to have been widespread and "not uncommon" in the 1830s. The species is no longer common, and its numbers are declining in northern portions of its range. With a global population of <15,000, the Andean Condor is the third least common of all New World vultures. The species was little studied until growing conservation concerns for the California Condor in the 1980s raised its profile as a surrogate, and Andean Condors are now the focus of many investigations.

The species is one of only a few sexually dimorphic scavenging raptors; specifically, the only one in which males are notably larger than females. Adult males have prominent dewlaps; thick, white feathered ruffs; gray-brown irises; and a bare, dark-red head with a large fleshy comb on its crown. Adult females lack the elaborate head festoons and have red irises. The sexes are otherwise similar in plumage. Genetics place the species in the second major clade in family Cathartidae, a three-species lineage that split from *Coragyps* and *Cathartes* approximately 14 million years ago. The species is the sole representative of a genus that split from the King Vulture and the California Condor more than 3 million years ago. Its fossil record dates from the late Pleistocene.

The species was described and named *Vultur gryphus* by Carolus Linnaeus in 1758, a binomial that it still retains. The genus name *Vultur* is Latin for "vulture"; and the species name *gryphus* is Greek for "hook-nosed." The vernacular name "condor" is from "cuntur," the Quechua (Inca) name.

Subspecies. A recent examination of geographically based genetic structure involving samples from across its distributional range suggests "moderate" genetic differentiation between individuals from northern and southern regions of the species range, suggesting site-specific, directional dispersal following natural updraft corridors along isolated mountaintops in the Andes.

Breeding biology. Andean Condors are obligate cliff-nesters whose breeding is restricted to locations with suitable inaccessible, cliff-nesting sites. Incubation of its single-egg clutch takes 54–58 days, and like many large-bodied raptors, protracted periods of nestling and fledgling care result in pairs producing young at intervals of 2 or more years. Although behavioral and physiological aspects of its reproduction have been studied in captive pairs, studies of breeding in the wild, including the aspects of both nest-guarding and chick-rearing are limited to observations of a single pair near Bariloche, Argentina, by Argentinian biologists Sergio Lambertucci and Orlando Mastrantuoni in 2005–2007.

Lambertucci and Mastrantuoni began the 28-month study in late January 2005 when two adults began roosting within several hundred meters of the eventual nesting site. They were observed most nights through March of that year. The pair roosted approximately 50 ft (15 m) from the eventual nest cave for the first time in mid-May, and nightly thereafter starting in early August. The pair copulated on a cliff above the nest site in early October, after which at least one individual remained at or in the cave continually. The pair successfully defended the site against both Turkey Vultures and Peregrine Falcons when incubating, and for 2 months after hatching. They continued to alternate time at the cave through January 2006, but gradually spent time outside of it, either flying about or perched at the entrance. The chick was first seen at the entrance in late January, when it was estimated to have been about 2 months old. An initial extended flight of 4 min occurred in early June when the young bird flew with both adults before perching 90 yd (80 m) from the cave entrance. Both adults continued to feed the juvenile daily through February 2007, typically in response to its begging. The chick was fed 102 times in January–February, with feedings occurring between 1100 and 1800 (11 am and 7 pm). The male spent more time near the nest site than did the female mate, and fed the young bird up to 4 times a day, compared to 2 times a day for the female. The juvenile was last seen at the site in early March 2007. Adults remained in the area through late April when observations ceased. Presumably the same pair was seen courting at the site in September and appeared to have initiated incubation in October.

Diet and feeding behavior. Early observers, including Charles Darwin, noted that condors fed on carcasses of both guanacos and rheas inland, and stable-isotope evidence indicates that condors once depended heavily on marine mammals coastally. Observations and regurgitation pellets collected in northern Patagonia in the 1990s indicate that condors now rely principally on livestock, including sheep and goats as well as introduced red deer and rabbits for more than 98% of its diet. In southern Patagonia, predator kills, including those of pumas, figure largely in the species' diet. The extent to which shifts to smaller food items influence food-searching behavior in condors has yet to be explored, but searching for small versus large carcasses most likely selects against searching in large groups. Recent increases in populations of Black Vultures also may affect food searching in the species, as field evidence indicates that groups of the Black Vultures deter the condors from feeding at large carcasses in much the same way that groups of *Gyps* sometimes deter Lappet-faced Vultures at large carcasses.

Other factors affecting condor feeding behavior include both significant sexual size dimorphism and age-related behavioral dominance. Male condors, which can be 30–40% more massive than females, dominate the latter at carcasses, and within sexes, older birds tend to dominate younger individuals. In high-quality mountainous areas where large numbers of condors congregate around carcasses, intraspecies dominance may result in subordinate individuals searching for carcasses in lower-quality, lowland areas, resulting in breeding females feeding later in the day, thereby increasing the cost of late afternoon soaring flights. That sex ratios, which are female-biased in juveniles and subadults, shift to being male-biased in adults, suggests a possible impact of sex-related differences in behavioral dominance.

Social behavior. The American ethologist Jerry McGahan's meticulous observations of the social behavior of Andean Condors at baited carcasses in Colombia and Peru in the late 1960s and early 1970s provide some of the most helpful insights into the social behavior of the species. Aside from the age- and sex-specific differences mentioned above, McGahan noted that adult pairs flew and perched in pairs far more often than their relative numbers suggested and that intraspecies aggression was relatively common at carcasses. He also noted that aggression at carcasses sometimes resulted in subordinate individuals regurgitating before forced takeoffs, and that regurgitated food was quickly consumed by individuals that remained. Reciprocal preening involving "mandibulations" of skin and feathers around the head and neck, resembling those between parents and their young, also occurred among paired and non-paired birds, as well as between adult pairs prior to copulation. Within pairs, males usually flew from a carcass or roost first, followed by the female. He also noted that adult males generally avoided each other, and that when they did come together, aggressive behavior often followed. Adult males participated in fights at carcasses more than other individuals, and adult male–adult female interactions and adult male–immature male interactions confirmed adult male dominance. He also noted that adult female–immature male interactions did not demonstrate clear dominance, and in such instances female dominance appeared to be reversed in older immature males. In addition, immature males seemed to be attracted to one another, yet remained aggressive toward one another, and that younger birds appeared more "hesitant" than older individuals to attempt supplantings at carcasses. Finally, in Colombia, adult pairs maintained regular and exclusive cliff-roosting areas, whereas in Peru, groups often roosted in groups, both during the day and at night.

Although none of the above is surprising, the consistency with which these events occurred suggests a high degree of social structure, at least in southern Colombia and coastal Peru.

Flight behavior. Strong updrafts are a more important element for successful low-cost soaring flight in large and more heavily wing-loaded vultures than in smaller and more lightly wing-loaded vultures. Like other large scavenging raptors, Andean Condors occur in areas where topographic relief provides relatively constant, strong, cliff-face and deep canyon updrafts for soaring and gliding flight. Efficient high-speed gliding also allows large-bodied species like Andean Condors to travel quickly among areas with predictable strong updrafts, and evidence shows that the species travels quickly when gliding. Flex-gliding, a flight behavior in which the wing area is reduced and the wingspan is sometimes halved by relaxing the patagium and bending the wrist, serves to speed the rate of cross-country travel in the species, often to the point that individuals glide faster than 90 mph (60 km/hr).

Andean Condors routinely use high-cost flapping flight, both when taking off from level ground and when engaged in high-speed aerial chases. The extent of their doing so was recently measured during 235 hours of observation in eight subadults fitted with flapping-rate data loggers. The study, conducted in Argentina by a team of six European and South America researchers, discovered that flapping flight was remarkably low, representing 1.3% of all flight time, interspersed by bouts of nonflapping gliding and soaring flight that ranged from 98 to 37 min per bird. Flapping during takeoffs was responsible for 75% of all powered flight daily, and that at least for Andean Condors, flapping flight was limited by energetic requirements associated with takeoffs. The degree to which this occurs in other species has yet to be quantified.

Movements. As is true for many sedentary vultures, Andean Condors have large home ranges. Twenty adults satellite tagged near Bariloche in northern Patagonia, Argentina, flew an average of 93 mi (150 km) daily while ranging across an area of more than 35,000 mi^2 (90,000 km^2) during the first 6 months post-release, with the largest-ranging birds covering an area more than 21,000 mi^2 (53,000 km^2). Nine of the birds flew into Chile where six of them bred. Although the species is generally considered non-migratory, at least some individuals routinely travel to areas distant from the Andes, with birds making seasonal, multi-thousand-kilometer round trips to feed as far east as the Rio Jauru in Mato Grosso, south-central Brazil, where in late May and early June they feed on the carcasses of cattle accumulated during the dry season on a "Vulture Island" oxbow together with Black Vultures and King Vultures. The degree to which such journeys affect survivorship and breeding ecology remains unexplored.

Conservation status. With a global population of <15,000, and with population declines and range retractions in the northern part of its range, the Andean Condor is currently considered **Near Threatened** by BirdLife International.

OLD WORLD VULTURES

Palm-nut Vulture *Gypohierax angolensis*

Spanish: Buitre Palmero
German: Palmgeier
French: Palmiste Africain
aka: Vulturine Fish-eagle

Taxonomy: Subfamily Gypaetinae

Size: Length 22–26 in (57–65 cm); wingspan 69–83 in (175–210 cm); 2.6–4 lb (1200–1800 g)
Estimated world population: 240,000; stable
Movements: Sedentary
Social ecology: Often gregarious; individual nesting; communal roosting; both individual and group food finding

The Palm-nut Vulture is an ecologically unusual and phylogenetically puzzling species. Relatively small and largely equatorial, Palm-nut Vultures have a 7 million mi^2 (18 million km^2) sub-Saharan range closely tied to the distribution of its dietary namesake, the nuts and endocarp of *Elaeis* and *Raphia* oil palms. The species may be the most sedentary of all Old World vultures.

Palm-nut Vultures are distinctive and arguably awkward birds with large and strong, laterally compressed, aquiline bills. Individuals have unfeathered red tarsi, and large red to reddish-orange, unfeathered eye patches. Adults have white and black plumage. Juveniles are feathered various shades of brown; and 2-to-4-year-old immatures are mottled whitish and dark brown. Often described as "unmistakable," and only sometimes confused with Egyptian Vultures and African Fish Eagles, the Palm-nut Vulture remains little studied overall despite its approachable nature, striking appearance, and the ease with which it can be approached in the field. The species global population has not been assessed since the early 1990s, when it was estimated at 240,000 individuals.

Palm-nut Vultures are one of the most tropical of all scavenging raptors, with most populations occurring no more than 10° north or south of the equator. A seemingly non-aggressive inhabitant of coastal mangrove swamps and wet inland savannas, Palm-nut Vultures occur at high densities with counts in some areas exceeding 50 individuals per linear kilometer.

The species was named *Falco angolensis* by the German naturalist and botanist Johann Friedrich Gmelin in 1788, based on a specimen likely collected near Luanda, Angola. The genus name changed several times before 1835, when German explorer Eduard Rüppell assigned it the monospecific *Gypohierax*. The common name vacillated between Palm-nut Vulture and Vulturine Fish Eagle well into the mid-20th century. The genus name *Gypohierax* is Greek for "vulture-hawk"; the species name *angolensis* is Latin for "of Angola."

Much of the confusion regarding the species' phylogenetic affinity arises from the fact that although Palm-nut Vultures exhibit several skeletal affinities with at least one Old World vulture, the species also has numerous non-vulturine characteristics, including habitually feeding on vegetable matter, routinely taking live fish, and perching in a relatively upright manner. Recent molecular phylogenies place Palm-nut Vultures together with Bearded Vultures, Egyptian Vultures, and Madagascar Serpent Eagles in the ancient vulturine subfamily Gypaetinae. Although these four species are genetically divergent, all are more closely related to each other than to Old World species in the subfamily Aegypiinae.

Subspecies. No subspecies are recognized.

Breeding biology. French raptor specialist Jean-Marc Thiollay studied the species' breeding behavior for 6 years in Ivory Coast in the early 1970s, reporting that individuals refurbished and reused nests across years, that there was little evidence of large-scale movements among adults, and that single-egg clutches were incubated by both adults for 44 days. Thiollay also reported that nest predation by primates and corvids appeared to be limited, with pairs fledging approximately 0.5 young per active nest. Although he reported that one fledgling remained with its parents for 6 months, Thiollay did not see other young with parents beyond 2 months post-fledging. Large-scale dispersal did not occur, and juveniles and immatures were often seen in breeding areas.

Diet and feeding behavior. Palm-nut Vultures are largely vegetarian feeders that live mainly on the nuts of African oil palms as well as upon the mesocarps of *Raphia* palms. Based on field observations, Thiollay estimated that individuals spent 60% of their time foraging on palm nuts, and 40% of their time preying on fish, crabs, and other marine invertebrates, which they took while flying low over wetlands. In addition, they routinely visited recently burned areas in search of exposed prey and carrion. The species also takes Grey Parrots in aerial pursuit, and they consume recently hatched green turtles. Palm fruit made up 65% of the gut contents of adult individuals and 92% of those of younger birds. That said, stable isotope analysis of regurgitation pellets collected in Poilão Marine National Park in coastal Guinea-Bissau identified crabs and oil palm fruits at frequencies of 69% and 34%, respectively. Although an early anecdotal report indicates that a single captive bird preferred palm fruit over meat, more recent cafeteria trials involving numerous free-ranging birds suggest that individuals select fish over palm fruits. How individuals deal with a diet high in plant material remains unstudied, but researchers estimate that a single bird could meet its daily caloric needs with as few as 34 fruits a day.

Social behavior. Little studied, but pairs appear to be territorial, although individuals often nest in close proximity and frequently interact socially. The species roosts communally, and occasionally with other species of vultures.

Movements. Largely sedentary, with adults remaining near nests year-round. Immatures may be more mobile but there are no indications of nursery areas as occur in *Gyps*. Palm-nut vultures have not yet been satellite tracked.

Conservation status. Given an apparent stable population of 240,000, its large range, and the overall availability of its wetland habitat, BirdLife International currently ranks the Palm-nut Vulture as a species of **Least Concern**.

Bearded Vulture *Gypaetus barbatus*

Spanish: Quebrantahuesos
German: Bartgeier
French: Gypaète barbu
aka: Lammergeier

Taxonomy: Subfamily Gypaetinae

Size: Length 37–49 in (94–125 cm); wingspan 91–112 in (230–285 cm); 10–16 lb (4800–7200 g)
Estimated world population: <30,000; recent rapid decline
Movements: Sedentary; immatures sometimes wide-ranging and nomadic
Social ecology: Solitary; territorial nesting; individual and group food finding

The Bearded Vulture is a massive and behaviorally distinct, mountain-dwelling species, with a highly disarticulated range of 24 million mi^2 (62 million km^2) in mountainous southern Europe, Africa, the Indian subcontinent, and the Tibetan Plateau. Unique in being the only bird that has an adult diet consisting largely of bones, this physically distinct species has a fully feathered head with a small black beard protruding from the base of its bill, and mainly blackish wings, back, tails, and facial masks. In most populations, the remaining whitish feathers of the head, neck, and body are purposely stained orange as a result of dust- and mud-bathing in wet, iron-rich soils, which is an apparently innate behavior that is most likely related to intraspecies signaling (see Box 2.3).

The species was named *Vultur barbatus* by the Swedish taxonomist and naturalist Carolus Linnaeus in 1788, based on a specimen collected in Algeria. The genus name was changed to *Gypaetus* in 1784. A second common name, Lammergeier, or "lamb vulture" in German, was widely used well into the 20th century, before falling out of favor due to the questionable reference to its preying on lambs. The genus name *Gypaetus* is Latinized Greek for "vulture-eagle"; the species name *barbatus* is Latin for "bearded."

Subspecies. Two subspecies, differing in both size and plumage, are currently recognized: the larger and more distinctively marked *barbatus* occurs in Europe, northwestern Africa, and Asia; the smaller and less distinctively marked *meridionalis* occurs elsewhere in Africa and southwestern Arabia

Breeding biology. Bearded Vultures have a relatively protracted breeding season of up to 10 months that varies considerably throughout its range, including December through September in Eurasia and North Africa, and May through January in southern Africa. Pairs build and subsequently refurbish large stick nests, usually in potholes, small caves, and other protected areas, where they lay single, or more commonly, two-egg clutches. Both parents incubate. Chicks take about 2 days to hatch. Parents bring food to the nest with their feet and feed their nestlings bill-to-bill. Meat is fed to nestlings and to young fledglings; bones are fed to older fledglings. In southern Africa, nestlings are left unattended at about 60 days of age and fledge at 124–128 days. Parents continue to feed young for 4 to 4.5 months after hatching.

Spanish vulture specialist Antoni Margalida and others carefully monitored the biology of 20 pairs of Bearded Vultures in the Pyrenees, and their findings offer some of the best details of breeding ecology in the species. Most individuals laid eggs in early January and averaged 43% nesting success, with early nests being more successful than later nests. Overall, both parents contributed significantly to nesting success, with males supplying most nesting material and providing most territorial defense. Almost half of all nests were usurped, some by Griffon Vultures. Pairs defended their nests against ravens and Golden Eagles, with the former suspected to be predators and kleptoparasites, and the latter to be nest-usurpers. Incubation lasted an average of 54 days and was shared equally between pairs, as was nest attendance and feeding. Two-egg clutches were laid in 80% of all nests, but mortality during hatching and the first several weeks post-hatching limited successful nests to a single fledgling. Nestling mortality was thought to be linked to a dietary dependence upon meat versus bones by hatchlings, and to siblicide, the latter being facilitated by the older nestling aggressively starving the younger. Adults carried bones to the nest in their talons or bill. Parental regurgitation was not observed, although parents sometimes "recycled" regurgitation pellets from nestlings. Nestlings left the nest at about 4 months of age, with males tending to fledge earlier than females.

Diet and feeding behavior. Bones usually make up 70–90% of the diet, making the species the only vertebrate to feed largely on this tissue. Individuals swallow bones measuring up to at least 10 by 1.5 in (25 by 3.0 cm), and they smash larger bones into smaller pieces at traditional rocky ossuaries where the bones are thrown, sometimes repeatedly, from heights of 60 ft (18 m) while flying downwind. Although the species' digestive tract is not notably long and lacks a specialized region for mechanically breaking bones, it features a high concentration of acid-producing epithelial cells that enhance rapid chemical decalcification. The species also possesses a highly elastic, thickly lined esophagus that stores bones during decalcification and provides protection from the ragged ends of broken bones.

Studies of captive individuals indicate a digestive efficiency of 75–80%. That efficiency together with the remarkably high energy content of bone, similar to that of mammalian muscle, results in vultures consuming a largely bone diet (with attached muscle and skin) and absorbing almost 90% as much energy per unit mass as one consuming a diet of viscera and muscle tissue. Field evidence suggests that Bearded Vultures prefer old to fresh bones, most likely because the former weigh less and, therefore, are less costly to transport. Individuals tend to choose "fattier" long bones, such as tibial and tarsal, which are high in oleic acid and thus more valuable nutritionally, over less-fatty mandibles, scapulae, and vertebrae. Skeletal material dehydrates more rapidly in non-tropical mountainous areas than does soft body tissue, which makes it quickly unavailable to insects and microbes. Bone-eating scavengers are able to survive on fewer carcasses than their meat-eating counterparts, enabling the species to exist in high mountain areas with lower ungulate densities than other avian scavengers. The specialized bone diet also enables the species to feed non-competitively with other vultures that eat less bone.

Social behavior. Bearded Vultures are long-lived, territorial breeders with delayed maturity that protect nesting sites against potential intra- and interspecies usurpers and kleptoparasites. Although typically monogamous, polyandrous trios also occur. In a growing population in the Pyrenees, unpaired breeding-age males were subordinates in such trios, usually in high-quality territories. Polyandry, which was first reported in the region in 1979, occurred in 12% of 56 nests in 1988, and 15% of 92 nests in 2000. Although the reproductive success of primary males may have been compromised by the formation of trios, secondary males were not evicted from territories by the original pair. Polyandrous trios also have been seen attending active nests in the Drakensberg mountain range of southern Africa, in Corsica, and the Italian Alps.

Movements. Adults are largely sedentary. Satellite tracking of recently fledged juveniles, immatures, as well as both nonbreeding and breeding adults, provides considerable information regarding the movement ecology of age groups in Europe and South Africa. Nine recently fledged juveniles tracked in the Spanish Pyrenees remained near natal nests or release sites for 2 months post-fledging while making exploratory flights of up to 34–37 mi (55–60 km). Parentally reared fledglings also made repeated trips to bone-breaking ossuaries, initially with parents but eventually alone. Fledglings departed natal areas at about 190 days after their initial flights, or at about the same time parents started incubating their next clutch. Dispersal movements ranged from 4.5–118 mi (7.5–197 km), within the Pyrenees. Captively reared juveniles in the Alps and in southern Spain dispersed farther, most likely because of genetic differences, densities of existing regional breeding populations, and food availability, or the lack of suitable nearby breeding areas. Post-fledging movements of individuals tracked in the Drakensberg mountains indicated increasingly larger ranging areas during the first 6 months post-fledging 425 mi^2 (1099 km^2) minimum convex polygons (MCPs) than in European birds from the northern subspecies 380–1760 mi^2 (978–4544 km^2) MCPs.

Conservation status. With the exception of northern Spain, where numbers have increased since the mid-1980s, Bearded Vultures are declining globally. Poisoning, both intentional and incidental, together with land-use change, disturbance at breeding sites, and collisions with power lines, are the main threats to the species. Bearded Vultures are not common anywhere. With a global population of fewer than 10,000 birds, including fewer than 7000 mature individuals, the species is listed as **Near Threatened** globally, by BirdLife International.

Egyptian Vulture *Neophron percnopterus*

Spanish: Alimoche Común
German: Schmutzgeier
French: Vautour percnoptère
aka: Scavenger Vulture, Pharaoh's Chicken

Taxonomy: Subfamily Gypaetinae

Size: Length 21–26 in (54–66 cm); wingspan 57–69 in (145–175 cm); 3.5–5.3 lb (1585–2400 g)
Estimated world population: <10,000; recent declines
Movements: Partial migrant; immatures nomadic
Social ecology: Largely gregarious; sometimes nests in loose colonies; communal roosting; group food finding

Egyptian Vultures are small, slender-billed, dusky-white-and-black, obligate scavenging raptors with long, relatively narrow, pointed wings and distinctive, wedge-shaped tails. With an enormous 23 million mi^2 (60 million km^2) dry-country distributional range stretching from southern Europe and northern Africa west across the Middle East and into southern Asia, the species is one of the most widespread of all Old World vultures. Well-studied in many parts of its European and north African range, the species has an eclectic diet that includes carrion of all kinds, waste from trash dumps and abattoirs, bird eggs, animal feces, small living prey including terrapins, and insects, the latter both in flight and when following plows in agricultural fields. The inclusion of ungulate feces in its diet supplies carotenoid pigments essential for the species' brightly colored head and yellow face. The Egyptian Vulture is one of only a few birds known to engage in tool use and to apply an external cosmetic to modify the color of its plumage (Box 2.3).

Box 2.3 Cosmetic Mud Bathing

Birds use colorful plumage in many ways. Depending on the species, plumage can indicate age, physical condition, and social status, as well as sexual attractiveness. With notable exceptions, including the seasonally ornate breeding plumages of herons and egrets, plumage itself is usually static; however, it can be modified. Twenty-nine species in 13 avian families, including hornbills, pigeons, pelicans, ibises, woodpeckers, and parrots, purposely stain their feathers. Some species, like cranes, do so to camouflage themselves while incubating, others to enhance their sexual attractiveness. Some use pigmented secretions from their uropygial- or oil-gland, or from their skin, to modify their feathers; others use talcum-like feather "frass" from continually disintegrating feather down. Still others, including both Egyptian Vultures and Bearded Vultures, use external substances, including limestone sediment rich in iron oxide, to purposely re-color their otherwise naturally white feathers. They routinely dip and roll their head, neck, chest, and back in dry and damp, reddish iron-rich sediments. In Bearded Vultures, worn feathers hold greater amounts of iron oxide than those that are less worn, with iron accumulating mainly along the axes and ends of shafts, barbs, and barbules of feathers.

The evolutionary origins of deliberate vulture feather staining remain unclear. Some scientists suggest that it evolved either accidently when the individuals dust bathed in sediments rich in iron oxide to clean their feathers, or purposefully to camouflage the birds, or to protect feathers from abrasive wear or from parasitic feather lice. None of these explanations stands careful scrutiny. Neither Bearded Vultures or Egyptian Vultures have known natural predators they need to camouflage themselves from, and neither need to sneak up on carrion. Unstained and stained feathers wear equally, and feather lice survive just as long in pure water as in water laced with iron oxide. Social signaling, if not responsible for the evolution of feather staining, appears most likely to be responsible for maintaining it.

Mud bathing has been studied in greater detail in Bearded Vultures, where it has been known for decades, than in Egyptian vultures, where it was discovered more recently. It appears to be both routine and innate in the former, as captively reared young raised in isolation feather bathe just as often as do wild-caught adults. Furthermore, mud bathing appears to be innate in Egyptian Vultures. The behavior is correlated with age in Bearded Vultures, with adult feathers being more colorful than those of juveniles and subadults, as well as with sex, with females being more intensely colored than males. Observations indicate that bathing in reddish, iron oxide–rich soil increases following rains that rinse away the color of stained feathers, suggesting that individuals are able to recognize the effects of their staining efforts, either visually via self-recognition or tactilely during preening. Researchers who study the behavior in Egyptian Vultures have noted that mud bathing occurs less secretively than in Bearded Vultures, and that both males and females engage in it equally, which leads them to conclude that it functions principally for individual recognition in pair bonding or during sexual conflicts in the species. That it occurs in two phylogenetically related Gypaetine species, but not in other families or subfamilies of Old and New World vultures, remains unexplained.

Mentioned in the Old Testament, Egyptian Vultures were considered symbolic of royalty in ancient Egypt, where the species was once protected as a street cleaner in towns and cities. More recently, it has been derided as unclean because it feeds upon both human and animal feces.

The species was named *Vultur percnopterus* by the Swedish taxonomist and naturalist Carolus Linnaeus in 1758, based on a specimen from Egypt. J.-C. Savigny changed the genus name to *Neophron* in 1809. The long-used vernacular Pharaoh's Chicken reflects its former abundance in villages and cities in lower Egypt. *Neophron* is a figure in Greek mythology whose mother, Timandra, had an affair with Aegypius. Neophron is said to have tricked Aegypius into sleeping with his own mother, Bulis, and Zeus is said to have adjudicated the situation by changing both Neophron and Aegypius into vultures. The species name *percnopterus* is Greek for "dusky-winged."

Subspecies. Three subspecies are recognized. The nominate *percnopterus* occurs in Europe, Africa, the Middle East, and Asia, except for peninsular India where *ginginianus* occurs. *Majorensis* occurs on the Canary Islands.

Breeding biology. The species is monogamous. Both pairs and nests often are maintained across years. In spring, pairs build a relatively large stick nest that is frequently festooned with bones, small-animal carcasses, dung, and assorted human trash. Although usually a solitary nester, it sometimes nests in loose colonies.

Two-egg clutches hatch asynchronously; however, unlike the closely related Bearded Vulture, siblicide is rare, and many pairs fledge two young. In northern Spain, where 34 reproductive attempts were monitored in 2003–2005, occupied territories fledged an average 0.9 young, with 1.2 young fledging per successful nest. Incubation averages 42 days, decidedly shorter than in most vultures. Observations reveal substantial differences in the growth pattern of first- versus second-hatched nestlings, with the latter being smaller at most stages of development than first-hatched siblings, and with many of the former dying within two weeks of hatching. Both parents feed the nestlings via regurgitation. Young fledge at 70–85 days of age. Most initial flights involve flapping and gliding, with soaring flights occurring only gradually, possibly because nestling primaries measure approximately 84% of adult length at the time of first flight. Although recently fledged young sometimes locate carrion on their own, young typically follow parents to feeding sites, most likely to learn social skills at carcasses where conspecifics feed together, as well as to learn how to locate feeding sites. The post-dependence period is brief, particularly in migratory populations, with young in northern Spain initiating migration at 89 to 113 days of age. Young are not known to migrate with their parents.

Diet and feeding behavior. A human commensal in most of its range, the Egyptian Vulture has an eclectic diet. In high-density areas, the species often assembles in dozens at large vertebrate carcasses, particularly in situations where larger and more dominant *Gyps* are absent, as well as by the hundreds at slaughter houses and trash dumps, in much the same manner as do Hooded Vultures in West Africa.

The species is an exceedingly opportunistic, and at times, un-vulturine feeder. In Africa it uses stones to open birds' eggs, especially those of ostriches. Described in detail by Jane Goodall and photographer Hugo van Lawick in the late 1960s, the phenomenon of stone throwing was first noted in the mid-19th century. Stone-throwing occurs when individuals purposely break the shells of ostriches and other bird eggs with the intent of consuming their contents. Individuals do so by picking up small stones in its bill, raising its head with its bill pointing upward, and then projecting the stone in the direction of the egg with a "forceful movement of the head and neck," and repeating the pattern until the shell is broken and the content consumed. Individuals also slam and break smaller and more manageable eggs of both Lesser Flamingos and White Pelicans, as well as by smacking them against solid surfaces. The species' preference for rounded stones (rather than more jagged rocks) is believed to have originated from such "egg-throwing" behavior. Recent observations of Egyptian Vultures successfully smashing and eating an abandoned Griffon Vulture egg at a colony in southern Spain suggests the behavior may be widespread. An additional incidence of Egyptian Vulture tool use in

Bulgaria, where an individual used a twig to twirl and gather loose wool to add to its nest, suggests that tool use is a species proclivity. Although tool use was once thought to be transmitted culturally, experiments with captive individuals reveal that eggshell breaking is an innate behavior that improves with practice.

Social behavior. Gregarious in both high-density, commensal, and in lower-density non-commensal populations, as well as on migration, individuals often search for food in pairs and in small groups. The species typically feeds on the perimeters of groups of larger vultures at carcasses. In northern Spain, Egyptian Vultures roost communally, mainly in large pines, with Griffon Vultures, Red Kites, Black Kites, and Common Ravens along the Ebro River. Numbers peak in late July after young have fledged.

The species is a human commensal in the Canary Islands, which it appears to have colonized more than 2500 years ago, most likely shortly after the arrival of goat-herding Berbers.

A landfill in Oman at the tip of the Arabian Peninsula, 3 mi (5 km) south of the capital of Muscat, supported a population of >450 Egyptian Vultures in November 2013, with rummaging adults outnumbering first-year and subadult birds 2-to-1. The species behaves similarly in Ethiopia, where individuals habitually roost along power-line pylons and communication towers near human settlements.

Movements. Egyptian Vultures are the second-most migratory of all vultures, with most European and many Asian breeders and their offspring overwintering in northern and central Africa. Almost all migrants enter Africa via the Strait of Gibraltar and Sicily, as well as via the Gulf of Suez at the northern end of the Red Sea and Bab al Mandeb at the southern. Significant autumn crossings occur at Gibraltar, where at least 3500 southbound migrants cross, and at the Gulf of Suez, where several thousand cross. As in Griffon Vultures, most migrants pass in the middle of the day on light tailwinds, typically in small groups of up to several dozen. Most first-year and older subadults oversummer in Africa, where, presumably, they fine-tune their group foraging and other social skills. Adults generally precede younger birds, particularly on return migration. In Israel, many migrants perform elliptical migrations with many returning individuals following a more westerly route in spring, when they are deflected northward by the Gulf of Suez, than in autumn when individuals follow more easterly routes through Sinai.

Several Egyptian Vultures have been tracked by satellite. Two French juveniles that began migration in late August crossed Gibraltar to West Africa shortly thereafter. A Bulgarian individual that began migration in late August to early September crossed Suez in September. All three migrated at least 2100 mi (3500 km) during 20–25 days, traveling an average of 48–78 mi (77–125 km) each day. The French birds overwintered in the Sahel in southern Mauritania, with one returning to northern Morocco, presumably en route to France, when it was 3 years old. The Bulgarian bird spent most of its time in southeastern Chad in the Sudano-Sahelian zone but also reached the Central African Republic. MCP wintering ranges of the two French birds averaged about 17 mi^2 (45 km^2).

Two adults tagged in Spain overwintered in southern Mauritania, northern Senegal, and northern Mali, with one returning to the same wintering grounds in its second year of tracking. Migration routes varied both between birds and for individual birds, between years. Both performed elliptical migrations in Africa while traveling inland southbound, and coastally and more westerly when northbound. Maximum daily travel distances ranged from 239–317 mi (398–529 km) in autumn, and from 206–414 mi (343–690 km) in spring, with daily travel distances averaging 147–203 mi (245–339 km) during outbound movements and 89–157 mi (149–262) during return movements.

Three non-breeding individuals trapped and tagged in Oman and Djibouti that were tracked for up to 622 days, ranged over large MCP areas of 4900–7800 mi^2 (12,600–20,200 km^2) while visiting rubbish dumps and abattoirs in search of food. On average, high-use areas covered by these individuals were similar to those of 16 tracked subadults and adults in the Middle East and East Africa but were considerably smaller than those used by European migrants overwintering in West Africa.

Conservation status. With the exception of stable-to-increasing populations in Socotra, the Canary Islands, and Oman, and growing populations in Iberia, where numbers have increased since the mid-1980s, the species is declining globally. Threats include human disturbance, lead poisoning, direct and secondary poisoning including diclofenac in southern Asia, electrocution by power lines, collisions with wind turbines, and reduced food availability. Because of rapid declines in India caused by the veterinary drug diclofenac, and longer-term declines in both Europe and Africa, BirdLife International currently lists the species as globally **Endangered.**

Hooded Vulture *Necrosyrtes monachus*

Spanish: Alimoche Sombrío
German: Kappengeier
French: Vautour charognod

Taxonomy: Subfamily Aegypiinae

Size: Length 21–26 in (54–66 cm); wingspan 59–71 in (150–180 cm); 3.0–5.7 lb (1400–2600 g).
Estimated world population: <200,000; decreasing
Movements: Sedentary; adults largely sedentary; immatures episodically nomadic
Social ecology: Solitary to gregarious; solitary and loose colonial nesting; solitary and communal roosting; individual and group food finding.

The Hooded Vulture is one of Africa's smallest vultures. This slender-billed species is also one of its most widespread, ranging across an enormous 9 million mi^2 (23 million km^2) range across all but densely forested sub-Saharan and southern-most Africa. Most numerous in West Africa, where it is a human commensal in rural, suburban, and urban landscapes, it is far less commensal south of the equator, where it occurs mainly in natural areas. Although elegant in the air, on the ground the species' smallish head, seemingly oversized wings, and shuffling gate make it appear scrawny and malnourished. That it engages in feces-eating, or coprophagy, often in human-dominated landscapes, makes the Hooded Vulture seem pitiful to many, and abhorrent to others.

Adults are dark brown overall with a bare face and upper throat that, when excited, blushes red. Its thighs and the back of its neck are covered in fluffy and cottony whitish down. Juveniles are browner and drabber overall.

The species was first described as *Cathartes monachus* in 1823 by the Dutch taxonomist and museum director Coenraad Jacob Temminck, based on a Senegalese specimen. The genus name was changed to *Necrosyrtes* by the German zoologist Constantin Wilhelm Lambert Gloger in 1841. *Necrosyrtes* is Greek for "corpse-puller"; the species name *monachus* is Latin for "hooded."

Subspecies. Two subspecies are recognized, although recent genetic evidence suggests the two may represent full species. The nominate taxa *monachus* is a gregarious human commensal that occurs in West Africa, east to western Sudan and northern Uganda. The slightly larger and far less gregarious *pileatus* subspecies occurs in Sudan, Ethiopia, and East Africa, south to Namibia, Botswana, and South Africa.

Breeding biology. Courtship often involves regurgitation feeding. Pairs lay one-egg clutches in small stick nests and engage in biparental care. Incubation is estimated at 48–54 days. The less-studied *monachus* nests in trees and on human-built structures in both human-dominated and more natural landscapes. Breeding success at 56 nests in Burkina Faso exceeded 50%, where egg and nestling poaching and removal of stick nests for cooking fires were principal threats. South of the equator, *pileatus* builds small stick nests in riparian areas, usually in the main fork of densely leafed trees, sometimes in loose colonies with nearest-nest distances averaging 0.47 mi (0.75 km). In Kenya and Zimbabwe where conspecific theft of nest materials occurs, nests can be reused for as many as 12 years. Breeding success at 12 *pileatus* nests was 50–70% during 2 years of observation on the Olifants River in northeastern South Africa, where baboons and Martial Eagles were known predators.

Parental care has not been studied in detail, but observations of satellite-tracked nestlings suggest it continues for several months post-fledging, with some fledglings remaining close to nests for up to 6 months.

Diet and feeding behavior. Commensals north of the equator visit both large and small abattoirs and trash dumps to feed on offal and butchery scraps, as well as on the carcasses of domestic and wild ungulates, and on insects, including swarming termites. The species also feeds on both feline and human feces, and follows African painted dogs, in part, to feed on their feces. When feeding at carcasses with larger vultures, Hooded Vultures peck on carcass debris at the perimeter of multi-species feeding assemblages. The species visits small roadside butcheries in West Africa, where it meanders among the legs of workers who pay little, if any, attention to them other than to offer small scraps. It also feeds on scraps left by hyenas and lions, both in the wild and in zoos. Individuals also take scraps left by "breakfast parties" at tourist balloon-landing sites in the Maasai Mara National Reserve in Kenya.

Social behavior. Little studied, although group bathing and drinking routinely occur near communal roosts.

Movements. Field evidence suggests that many populations are sedentary. Exceptions include several West African populations that move into and out of the Sahel seasonally, and those in Sudan, which leave savanna areas following the rainy season. Satellite-tracked subadults sometimes disperse in episodic bouts of nomadic behavior. Kernel density estimation (KDE) of home-range densities of satellite-tagged and tracked individuals in eastern and southern Africa were 30 and 160 times larger, respectively, than those of commensals tagged in West Africa, most likely because of the availability of human-subsidized food there. Males have larger home ranges than do females, both within and outside of the breeding season. As in other vultures, juveniles have a larger monthly home range than do adults.

Conservation status. With populations exceeding 15 individuals per linear km on roadside counts in The Gambia and other parts of West Africa, the Hooded Vulture is the most common species in human-dominated landscapes regionally. Numbers appear to be declining in many portions of its range due to indiscriminate poisoning, modern sanitation, power-line collisions and electrocutions, as well as to their use in witchcraft and as bushmeat. Because of an estimated global population of fewer than 200,000 individuals, and rapid population declines in many parts of its range, BirdLife International currently lists the species as **Critically Endangered** globally.

Indian Vulture *Gyps indicus*

Spanish: Buitre indio
German: Indiengeier
French: Vautour indien
aka: Long-billed Vulture

Taxonomy: Subfamily Aegypiinae

Size: Length 32–41 in (81–103 cm); wingspan 77–102 in (196–258 cm); 12–14 lb (5500–6300 g)
Estimated world population: <30,000; recent rapid decline
Movements: Sedentary; immatures somewhat nomadic
Social ecology: Gregarious; communal nesting and roosting; group food finding

With a historical global population in the multiple millions and a distributional range of 0.8 million mi^2 (2.1 million km^2) stretching from Nepal, west into southeastern Pakistan, and south into much of peninsular India, this medium-sized *Gyps* was once considered one of the world's most common large-bodied raptors. Catastrophic population declines that began in the 1990s decreased the species' range considerably and reduced its current global population by more than 90%. Unfortunately, the species was little studied prior to the declines, and knowledge of its previous ecology is limited. What is known suggests that the species' scavenging behavior differs somewhat from that of African *Gyps* in being more closely associated with, and dependent on, human-dominated landscapes, including cities, towns, and villages, as well as cultivated areas and woodlands. The extent to which the species' former ecology will emerge as populations recover from recent declines remains to be seen. Awaiting study is a careful investigation of its ecological relationships with two regional congeneric counterparts, the closely related and slightly larger Slender-billed Vulture and the somewhat more phylogenetically distinct and slightly smaller White-rumped Vulture.

The Indian, or once called Long-billed Vulture, was named *Vultur indicus* by the Austrian naturalist Giovanni Antonio Scopoli in 1786, based on a specimen collected in India. The genus name was changed to *Gyps* shortly thereafter. In 1849, the American ornithologist John Cassin perhaps said it best when he characterized the species' taxonomy as being "in an extraordinary state of confusion"—a status it retained until recently. For more than a century the Indian Vulture was lumped together with the Slender-billed Vulture, with the latter being considered a subspecies and listed as *G. i. tenuirostris* or *G. i. nudiceps*. A comprehensive study in 2001 by the American ornithologist Pamela Rasmussen split the two forms into separate species, an idea originally suggested by the British naturalist G. R. Gray in 1844. The Indian Vulture differs in many ways from the Slender-billed Vulture, specifically, by having more elongate, narrow nares; a paler, bluish gape and eye ring; a pale yellowish bill and cere; a thicker neck; shorter, straighter, paler, and rounder-tipped feathers of the breast, mantle, and wing coverts; fuller and whiter neck ruff feathers; and covered, but not necessarily visible, downy white thighs.

The genus name *Gyps* is Greek for "vulture"; the species name *indicus* is Latin for "of India."

Subspecies. No subspecies are recognized.

Breeding biology. The species is little studied overall. It nests principally in December and January, usually in colonies of one to two dozen pairs, and lays single-egg clutches in stick nests mainly on cliff ledges, rocky outcrops, and buildings, and less commonly in trees.

Diet and feeding behavior. Again, little studied overall. Early accounts often lumped the species with Slender-billed Vultures, White-rumped Vultures, Himalayan Vultures, and Griffon Vultures. The species feeds together with other *Gyps* on livestock and wild ungulate carcasses, where it is dominated by the larger Griffon and Himalayan *Gyps* and Cinereous Vultures, but not by White-rumped Vultures.

One of the most informative investigations of the species is that of the Indian ornithologist R. B. Grubh, who studied it together with White-rumped Vultures and lesser numbers of Griffon Vultures in the early 1970s for his PhD work. His study included the 490 mi^2 (1265 km^2) Gir Wildlife Sanctuary for Asiatic lions in west-central India, at a time when all three vulture species were still common. Grubh reported that the three species fed mainly on the carcasses of water buffalo and Zebu cattle that had been killed by lions or that had died of natural causes. Grubh's observations of feeding behavior at 194 carcasses across 3 years indicated that interspecies interactions at group feedings were determined by size, with the smallest species sometimes using gang behavior at carcasses to overwhelm the actions of the two larger species. Grubh also noted that individuals were particularly wary of carcasses at which Asiatic lions were seen nearby, as well as when surrounding vegetation offered little in the way of takeoff space for rapid flight. He estimated that approximately 440 vultures used the sanctuary and that food availability probably was sufficient to supply all of the species needs. He also noted that soaring vultures sometimes "led" lions to carcasses, and that during moonlit nights carcasses often were fed upon well into the night. Individuals approached lion kills suspiciously, and even then, only after the lion or lions had moved well away. Grubh's observations led him to conclude that competition between lions and vultures was small, and that human presence near carcasses, which tended to drive away lions, most likely benefited vultures.

The projected increases in numbers of recently decimated populations of Indian Vultures and other *Gyps* as they recover from their catastrophic declines will provide ecologists with opportunities to study them more fully. Specifically, investigations will be intriguing that assess the geography of recolonization, both inside and beyond so-called vulture-safe zones (see Chapter 7 for details), and that explore how interspecies competition, including gang behavior, affects reproductive success and population growth among the species involved.

Social behavior. Little studied except for the Grubh observations mentioned above.

Movements. Little studied.

Conservation status. The continental loss of arguably the three most common southern Asian vultures due to poisoning by the veterinary drug diclofenac, including the now **Critically Endangered** Indian or Long-billed Vulture, offers both numerous opportunities and challenges for conservation biologists. Lessons from successful vulture reintroduction efforts in the Americas, Africa, and Europe should be borrowed upon rigorously. Adaptive management and external oversight will be critical to success, as will incorporating extensive monitoring and observational efforts.

Slender-billed Vulture *Gyps tenuirostris*

Spanish: Buitre picofino
German: Dünnschnabelgeier
French: Vautour á long bec

Taxonomy: Subfamily Aegypiinae

Size: Length 30–40 in (77–103 cm); wingspan 77–102 in (196–258) cm; mass, no data
Estimated world population: <2,500; declining
Movements: Largely sedentary; immatures sometimes nomadic
Social ecology: Solitary to highly gregarious; solitary and loose colonial nesting; communal roosting; individual and group food finding

The Slender-billed Vulture is a small- to medium-sized Asian *Gyps*, which until recently was considered a subspecies of the Indian or Long-billed Vulture. Because of its cryptic taxonomy, many aspects of the basic biology of the species remains relatively unstudied. The species breeds in forested areas across more than 770,000 mi^2 (2 million km^2) of the Gangetic Plain and lower foothills of the Himalayas, into parts of Southeast Asia.

The British naturalist and founder of Indian ornithology, G. R. Gray, named and correctly identified the species as *Gyps tenuirostris* in 1844, based on a specimen collected in Nepal by Brian Hodgson. Shortly thereafter, however, the species was lumped as a subspecies with the Indian Vulture, sometimes as *G. i. tenuirostris* or, alternatively, as *G. i. nudiceps.* A comprehensive study published in 2001 by the American ornithologist Pamela Rasmussen supports its identity as a full species, as originally suggested by Gray.

The species differs from the Indian Vulture in having a more slender and completely bare head and neck, and a pale culmen ridge on its otherwise dark bill (versus the yellowish bill of the Indian Vulture), and a conspicuously thinner neck and decidedly less conspicuous bare pectoral patches, among other features including considerable plumage differences. Extensive field studies of the species have yet to be conducted.

The genus name *Gyps* is Greek for "vulture"; the species name *tenuirostris* is Latin for "slender-billed."

Subspecies. Insufficient data. No indication of geographic variation.

Breeding biology. Unlike the closely related colonial cliff-nesting Indian Vulture, the Slender-billed Vulture nests only in trees, usually 8–27 yd (7–25 m) up, and principally alone, although it sometimes nests in loose colonies. It lays a single-egg clutch, the incubation of which is unstudied.

Diet and feeding behavior. Exclusively large carrion, including feeding on a leopard-killed carcass by moonlight.

Social behavior. Not well studied, but said to roost in loose colonies.

Movements. Largely sedentary, although some individuals move south in winter.

Conservation status. Declines due to diclofenac, perhaps not as severe as in White-rumped Vultures, occurred in the 1990s. Because of recent declines, a small remnant population, and the threat of continued diclofenac poisoning in parts of its range, the species is listed as **Critically Endangered** by BirdLife International

White-rumped Vulture *Gyps bengalensis*

Spanish: Buitre Dorsiblanco Bengali
German: Bengalengeier
French: Vautour chaugoun

Taxonomy: Subfamily Aegypiinae

Size: Length 30–34 in (75–85 cm); wingspan 76–84 in (192–213 cm); 8–13 lb (3500–5700 g)
Estimated world population: <10,000; recent declines
Movements: Largely sedentary; immatures sometimes nomadic
Social ecology: Largely gregarious; loose communal nesting and communal roosting; group food finding

The least massive of all *Gyps*, the White-rumped Vulture has a distributional range of 2.7 million mi² (7 million km²) that stretches from the Malaysian Peninsula northwest through India, China, Nepal, Pakistan, Afghanistan, and into southeastern Iran. Until the 1980s, the species ranked as one of the most common of all Old World vultures. Catastrophic population declines resulting from poisoning by the veterinary drug diclofenac in the 1990s brought a world population once estimated to exceed 10 million to fewer than 10,000 in the first decade of the 2000s. As in the Indian Vulture, the species was little studied prior to its precipitous decline, and knowledge of its pre-decline ecology is limited. The species behavior differs somewhat from that of African *Gyps* in its being a human commensal in many parts of its range. The extent to which its tolerance of humans will continue as the species recovers from recent population declines remains to be seen.

The species was named *Gyps bengalensis* by the German naturalist Johann Friedrich Gmelin in 1788, based on a specimen collected in eastern India. The genus name was changed to *Pseudogyps* in the first half of the 20th century, based on its relatively small size and its 12 as opposed to 14 tail feathers, a trait typical of larger *Gyps*. Recent molecular genetics, however, clearly place the species within the genus *Gyps*.

The genus name *Gyps* is Greek for "vulture"; the species name *bengalensis* is Latin for "of Bengal."

Subspecies. No subspecies are recognized.

Breeding biology. The species was little studied prior to its recent population decline. What follows summarizes what information is available.

The White-rumped Vulture is a largely colonial, albeit sometimes solitary, tree-nesting species that builds relatively large stick nests, 32–98 ft (10–30 m) high in tall, often evergreen, trees along watercourses and in villages. White-rumped Vultures typically nest between September and June with 2 to 15 nests in a single tree in loose colonies of up to 40 or more pairs. Single-egg clutches are typical, although two-egg clutches also occur. Biparental incubation is estimated at 45–56 days. Post-fledging care is unstudied. Dispersal movements by young birds have been reported. Whether immatures tend to assemble in nursery areas, as occur in African *Gyps*, remains unstudied.

Prior to their population collapse, many considered the White-rumped Vulture to be the most abundant large raptor globally, as well as the most common and widespread large species of raptor in India. Estimates suggest that at least 95% of the population disappeared between 1990 and approximately 2005. Unfortunately, significant population monitoring began only after the population collapsed, making it difficult to assess earlier numbers. Pre-decline studies in Keoladeo National Park (11.5 mi^2 [29 km^2]), 112 mi (180 km) south of Delhi, indicate 353 pairs (a density of 4.7 pairs per mi^2 [12 per km^2]) in the mid-1980s, with a nesting success of 82%. An urban population in Delhi in the late 1960s was estimated at approximately 400 pairs, or a density of 1 pair per mi^2 (2.7 per km^2).

Diet and feeding behavior. Like many other *Gyps*, the species depends heavily on large carcasses, particularly those of domestic livestock, and on human organic waste. As in other small *Gyps*, individuals are dominated by larger species one-on-one at carcasses, whereas large feeding groups can overwhelm smaller numbers of large species.

Social behavior. Largely unstudied except for the Grubh observations mentioned in the Indian Vulture account.

Movements. Little studied.

Conservation status. The continental loss of arguably the most three common *Gyps* species in the region, including the now **Critically Endangered** White-rumped Vulture, offers both numerous opportunities and challenges for conservation biologists. Reintroduction of wild populations should borrow heavily from the reintroduction efforts involving the California Condor. Adaptive management and external oversight will be critical to success, as will incorporating extensive monitoring and observational efforts.

Griffon Vulture *Gyps fulvus*

Spanish: Buitre Leonardo
German: Gänsegeier
French: Vautour fauve

Taxonomy: Subfamily Aegypiinae

Size: Length 37–43 in (93–110 cm); wingspan 53–106 (135–270 cm); 14–25 lb (6200–11,300 g)
Estimated world population: 650,000–690,000; increasing overall
Movements: Sedentary; adults largely sedentary; immatures nomadic-migratory
Social ecology: Solitary to gregarious; communal nesting and roosting; individual and group food finding

The Griffon Vulture, also known as the Eurasian Griffon, is a large, bulky *Gyps*, second in mass only to the closely related Himalayan Vulture. With a distribution of 8 million mi^2 (20 million km^2) stretching across several dozen countries from southwestern-most Europe and parts of northern Africa, east to Central Asia, the global population is approaching or exceeding 690,000 birds, with a European population of more than 60,000 birds. The species ranks as the most "successful" of all Old World Vultures and is also one of the more comprehensively studied.

The species was described by Carl Ludwig von Hablizl, a Prussian-born Russian botanist, in 1783, based on a specimen collected in western Iran. The genus name was changed from *Vultur* to *Gyps* in the early 19th century. Recent molecular genetics place the Griffon Vulture in a sister group with Rüppell's Vulture.

The genus name *Gyps* is Greek for "vulture"; the species name *fulvus* is Latin for "tawny."

Subspecies: The Griffon Vulture is one of only two *Gyps* with recognized subspecies. The nominate *fulvus* occurs range-wide except in the Himalayas of Pakistan and India; *fulvescens*, which occurs in the Himalayas, exhibits relatively low genetic divergence from *Gyps himalayensis*.

Breeding biology. The breeding biology of the Griffon Vulture is typical of cliff-nesting, colonial *Gyps* species. Most colonies consist of from less than a dozen to several dozen pairs, although some exceed 100 pairs. Nests, which are built of sticks with or without herbaceous linings, are refurbished annually, average 24–39 in (0.6–1.0 m) in diameter. In growing populations, individuals usurp previously used and, sometimes, active nests of other raptors including Egyptian Vultures, Bearded Vultures, Golden Eagles, and Bonelli's Eagles. Clutches, which almost always are single-egg, are incubated by both sexes for 53–57 days. Nestlings fledge at approximately 120 days and disperse several months later. Breeding success ranges from 30 to 80% per active laying pair, with about half of all failures occurring before hatching. Copulations involving individually marked birds in Spain indicate that females in subadult plumage are far more likely to breed than are subadult males.

Breeding numbers in Spain increased threefold in the 1990s and continue to increase into the first two decades of the 21st century, apparently in response to the increased availability of livestock carcasses and reduced human persecution.

Diet and feeding behavior. Like other *Gyps*, the Griffon Vulture is a large-carcass carrion eater that uses conspecifics and other vulture species to help find carcasses via an aerial information network. The species most often feeds upon the carcasses of domestic livestock and wild ungulates. In the densely populated Iberian Peninsula, some individuals specialize in taking live rabbits, and groups sometimes take vulnerable livestock. In areas where transhumance is practiced, Griffon Vultures follow livestock moving between upland and lowland grazing areas to take advantage of carcasses that may be left after the passage.

Social behavior. Griffon Vultures are colonial cliff-nesters whose colonies sometimes exceed 100 breeding pairs. Breeding adults feed both singly and in groups, usually within 24–48 mi (40–80 km) of their colony sites, with colony-based aerial information networks helping individuals locate carcasses.

Movements. Movements are relatively well studied, especially in the Mediterranean basin where large numbers of ringed, wing-tagged, and, most recently, satellite-tracked individuals have been followed. Post-fledging dispersal typically involves movements to so-called nursery areas, where young birds face less competition from adults.

Re-sightings and the retrieval of the carcasses of more than 100 individuals ringed and wing-tagged as nestlings in the 1990s on the Adriatic island of Cres in Croatia, indicate seasonal post-fledging movements into the Iberian and Balkan peninsulas, with some individuals migrating into northern Africa. Juveniles leave Cres soon after fledging, with at least some returning to the island 5 to 6 years later. Multiple re-sightings of wing-tagged birds indicate similar geographic movements across years.

Seasonal movements of largely immature Griffon Vultures migrating between Europe and Africa have been reported at the head of the Gulf of Suez in the eastern Mediterranean, and the Strait of Gibraltar in the west.

Counts at the Strait of Gibraltar, which average 4000 birds, are tenfold higher than at the Gulf of Suez, most likely because 80% of the European population breeds in Spain. One adult fitted with a satellite-tracking device in southern Spain in early 2013 completed outbound and return journeys to Senegal (1865 and 1678 mi [3000 and 2700 km]) in about 2 weeks, traveling at a rate of 22–27 mi^{hr} (37–45 km^{hr}), with maximum daily distances of 124 and 130 mi (206 and 216 km), respectively.

Observations of immatures flying south across the Strait of Gibraltar suggest that the water crossing significantly impedes the outbound migration into Africa, with many individuals repeatedly attempting and aborting passages for weeks until strong northern and western winds enable successful midday crossings of flocks of migrants.

Conservation status. Several populations are now threatened by the veterinary drug diclofenac, poison-laced carcasses targeting mammalian carnivores, lead pellets, and reduced food availability. Collisions with wind turbines and power lines are also problematic. Given its large distribution and overall growing populations, BirdLife International lists the Griffon Vulture as a species of **Least Concern** globally.

Rüppell's Vulture *Gyps rueppellii*

Spanish: Buitre Moteado
German: Sperbergeier
French: Vautour de Rüppell

Taxonomy: Subfamily Aegypiinae

Size: Length 33–38 in (85–97 cm); wingspan 89–100 (226–255 cm); 15–20 lb (6800–9000 g)
Estimated world population: <22,000; declining in portions of its range
Movements: A largely sedentary but wide-ranging sometimes partial migrant
Social ecology: Gregarious; communal nesting and roosting; group food finding

Rüppell's Vulture, the second largest African *Gyps* species, has a relatively large sub-Saharan range of more than 5.4 million mi^2 (14 million km^2).

The species was one of the last African vultures to be described and remains one of the least studied.

A large and robust vulture with high wing loading, the Rüppell's Vulture is a colonial cliff-nester throughout most of its range. Breeding biology, the seasonality of which frequently varies among years, has been best studied in Kenya and Tanzania, where it depends heavily on the carcasses of migratory ungulates. Considered non-migratory by many, observations of individuals routinely crossing the Strait of Gibraltar in both spring and autumn, along with long-distance movements within Africa, suggest the likelihood of partial migration.

The species was named *Gyps rueppellii* by the German zoologist Alfred Brehm in 1852, based on a specimen collected near Khartoum, Sudan. Eduard Rüppell, the German naturalist and explorer who first mentioned the species, had earlier misidentified it as a Griffon Vulture. The genus name *Gyps* is Greek for "vulture": the species name *rueppellii* is after Eduard Rüppell.

Subspecies. Rüppell's Vulture is one of only two *Gyps* with recognized subspecies. The nominate *rueppellii* occurs throughout its sub-Saharan range in West and East Africa. The smaller and paler *erlangeri* is limited to Ethiopia and Somalia.

Breeding biology. Rüppell's Vultures are cliff-nesters whose colonies range in size from less than several dozen to hundreds of pairs. A cluster of several colonies in the Gol Mountains of Tanzania was estimated to host more than 1000 nests during the late 1960s and early 1970s. Tree-nesting has been verified in southern Niger and in Senegambia, suggesting that the behavior is not as rare as once thought. Tree-nesting sometimes occurs near seemingly suitable cliff-nest sites, and the propensity for the phenomenon has yet to be studied in detail.

Fieldwork in Tanzania and Kenya indicates the timing of peak breeding can shift among and within years, and that bimodal peaks in breeding activity can occur at single colonies. The timing of breeding is associated with peaks in numbers of wildebeest carcasses at the end of the dry season and then again several months later during wildebeest calving. At one closely watched colony in Tanzania, where two distinct areas of the colony existed, one area, with favorable winds and morning thermals, and overhanging rocks that shaded nests from afternoon sunlight, laid eggs 2 to 3 weeks earlier than a second area that had less favorable morning thermals and greater exposure to the afternoon sun.

Diet and feeding behavior. At a well-studied site in East Africa, the species depends heavily on the carcasses of migratory ungulates. Recent satellite-tracking studies confirm that individuals preferentially cluster near migratory ungulates during the dry season, when herds experience high mortality.

Social behavior. Like other colonial cliff-nesters, Rüppell's Vultures use aerial information networks to locate carcasses locally and regionally. Observations of breeding adults suggest that individuals do not engage in protracted parental care of fledglings, raising the question of how immatures learn to find food on their own. Field evidence indicates that recently fledged Rüppell's Vultures spend time in nursery areas away from nesting colonies and other places that have high densities of adults in much the same manner as do the young of other Africa *Gyps*. Whether they do so in social groups is unknown. What is known is that larger *Gyps* dominate smaller species at carcasses, and that older individuals dominate younger birds. Whether such advantages result in immature Rüppell's Vultures dominating adult White-backed Vultures, or vice versa, remains unstudied.

Movements. Although the species is thought to be largely sedentary, individuals that were satellite tracked by the American vulture biologist Corinne Kendall reveal large-scale ranging behavior including protracted movements between a breeding colony in southern Kenya and a feeding area in southern Sudan. This, together with the facts that the species has a specialized hemoglobin alpha^D subunit with a high oxygen affinity that allows individuals to absorb oxygen at low partial pressures, and that an individual collided with a commercial aircraft flying at 36,000 ft (11,000 m) in coastal West Africa, suggests that exceptional high-altitude flight occurs in the species. If so, the species would be pre-adapted for long-distance migratory flights, which could explain the appearance of a flock of 43 immature Rüppell's Vultures on the Cap Blanc Peninsula near the Moroccan–Mauritanian border in late summer 1978, as well as the increasing number of individuals shuttling between the Strait of Gibraltar in spring and autumn, and the newly confirmed reports of the species breeding in Iberia.

Conservation status. Major threats include land-use change, declining food availability, and more recently, both intentional and accidental poisoning. With a declining global population of 22,000 birds, BirdLife International lists the species as **Critically Endangered** globally.

Himalayan Vulture *Gyps himalayensis*

Spanish: Buitre del Himalaya
German: Schneegeier
French: Vautour de l'Himalaya

Taxonomy: Subfamily Aegypiinae

Size: Length 33–38 in (85–97 cm); wingspan 89–100in (226–255 cm); 17.6–26.4 lb (8000–12000 g)
Estimated world population: <35,000; possibly decreasing
Movements: Largely sedentary; sometime altitudinal migrant
Social ecology: Small colony nesting, communal and individual roosting, individual and group food finding

A massive *Gyps*, and one of the largest of all Old World species, this condor-sized vulture ranges for more than 2.3 million mi^2 (6 million km^2) across the Tibetan Plateau of southwestern China, into Bhutan, Nepal, India, Kazakhstan, Uzbekistan, Kyrgyzstan, Tajikistan, Afghanistan, Pakistan, and Mongolia. A solitary and small-colony nester, Himalayan Vultures search for carrion in small groups above meadows and alpine shrublands, where they typically feed on the carcasses of humans, yaks, and other ungulates. One of the least studied *Gyps*, the Himalayan Vulture appears to have been affected less by veterinary drug diclofenac than were other more southerly distributed species, most likely because of its wild-ungulate, mountain-based prey base.

The species was named *Gyps himalayensis* by the Scottish ornithologist Allan Octavian Hume in 1869, based on a specimen collected in northern India. The genus name *Gyps* is Greek for "vulture"; the species name *himalayensis* is Latin for "of the Himalayas."

Subspecies. No subspecies are recognized. However, a molecular phylogeny indicates low genetic divergence between it and the Griffon Vulture subspecies *Gyps fulvus fulvescens*, with which it co-occurs in the Himalayas.

Breeding biology. Little studied. Pairs nest singly or in small colonies of up to six pairs from December through June in bulky single-egg stick nests. Nests in the Tianshan Mountains of northwestern China were constructed largely of reeds on south-facing slopes.

Diet and feeding behavior. The species is common near monasteries where both wild gazelles and domestic livestock are abundant, but they are not necessarily near villages. On the Tibetan Plateau, researchers estimate that domestic yaks make up 64% of the diet, human corpses about 2% (Box 2.4), and wild ungulates about 1%. Feeding Himalayan Vultures dominate smaller *Gyps*, but not the larger Cinereous Vulture.

Box 2.4 Sky Burials

When it comes to the kinds of carrion vultures feed on, public attention appears to focus on whether vultures eat each other and whether they eat dead humans. The short answer to both is yes. Vulture cannibalism is described in Chapter 4. Here I detail the human sky burials, and the role vultures play in them.

Known as charnel sites to Buddhists, the ancient funerary involves placing an exposed, and sometimes butchered, human corpse on a mountaintop to decompose. The body is either covered with animal skin or cloth or, in central Asia, exposed to the environment, for the purpose of it being eaten by facultative or obligate scavenging birds. The similar Zoroastrian practice involves the use of stone structures called *Dakhma*. Sky burials occur in the Chinese autonomous regions of Tibet, Qinghai, Sichuan, and Nei Mongol, as well as in Mongolia, Bhutan, and in Sikkim and Zanskar, India. The practice likely results from practical considerations. Sky burials typically occur in mountainous areas above the tree line where cremation is difficult, and in less mountainous locations where the soil is rocky and difficult to excavate.

The practice typically involves wrapping the body in white cloth and placing it in the corner of the house for 3 to 5 days, after which monks burn incense, read from scripture, and chant. Family members then remove clothing from the body and arrange it in the fetal position. A burial master may cut the body into parts. The vultures, sometimes attracted by smoke, are offered the "whole body," including internal organs. When the flesh has been stripped and only the skeleton remains, the bones are broken into pieces with mallets and the pieces offered to smaller scavenging birds.

Social behavior. Not studied.

Movements. Adults are largely sedentary, although wide-ranging. Recently fledged juveniles and immatures initially disperse and thereafter undertake short-distance seasonal altitudinal and latitudinal migratory movements. Eighteen immatures tagged with GPS bio-loggers in Bhutan and tracked for more than a year engaged in both latitudinal and altitudinal migrations with moves from high altitudes in Bhutan to significantly lower altitudes in Bhutan, India, Nepal, eastern China, and Mongolia. Tracked individuals traveled northward in April and June and reached summering areas in July. Southbound movements occurred in October and November. Migrants flew principally along river valleys, but they traveled at altitudes of more than 24,500 ft (7500 m) above sea level (asl) while crossing the Himalayas. Northbound movements were more rapid than southbound movements, with one bird traveling briefly at 88 mi^{hr} (147 km^{hr}). Individuals settled in landscapes with higher productivity in summer than in winter, typically in grasslands and meadows in the Tibetan Plateau and in steppe grasslands in Mongolia. Five of the birds died within a year of being tagged. Those that survived traveled more directly to summering areas, were less likely to fly into northern headwinds, and settled in less-populated areas than non-survivors. Bio-loggers revealed that tracked individuals used thermal and orographic updrafts, both when migrating and when searching for carrion.

Conservation status. Numbers of Himalayan Vultures in the Annapurna Conservation Area of northern Nepal declined by 67–70% in 2002–2005, with active nests declining by 84%. BirdLife International lists the species as **Near Threatened** globally, mainly because of the threat of diclofenac poisoning.

White-backed Vulture *Gyps africanus*

Spanish: Buitre dorsiblancoAfricano
German: Weißrückengeier
French: Vautour African

Taxonomy: Subfamily Aegypiinae

Size: Length 31–35 in (78–90 cm); wingspan 78–90 in (197–229 cm); 9.2–15.8 lb (4200–7200 g)

Estimated world population: 270,000; severe declines
Movements: Largely sedentary; immatures sometimes nomadic
Social ecology: Solitary to highly gregarious; communal roosting; solitary and loose colonial nesting; individual and group food finding

The White-backed Vulture is the smallest and most widely distributed African *Gyps*. Despite a distributional range of more than 9 million mi^2 (23 million km^2) across sub-Saharan Africa, the species received relatively little scientific attention prior to catastrophic declines in many parts of its range in both West and East Africa, and in parts of southern Africa. The species is particularly gregarious at carcasses, where it often outnumbers other species. The species nests singly or in loose colonies in small wooded areas, often near water.

A large-carcass, carrion-eating species that feeds mainly on the meat and organs of medium and large ungulates as well as on livestock, African White-backed Vultures find carcasses via a social network of both conspecifics and other species of vultures. Thought to be largely sedentary as adults, immatures disperse from breeding areas to spend time in nursery areas with low adult densities.

The species, which was the last African vulture to be recognized by science, was named *Gyps africanus* by Italian museum collector Count Tommaso Salvadori in 1865, based on a specimen collected near Sennar, Sudan. Like several other small *Gyps*, the genus name was changed to *Pseudogyps* based on having 12 rather than 14 rectrices, or tail feathers, before being reassigned to *Gyps* in the 20th century. Although it loosely resembles the White-rumped Vulture of southern Asia, a recent genetic phylogeny does not support a close relationship within the genus between the two. The genus name *Gyps* is Greek for "vulture"; the species name *africanus* is Latin for "of Africa."

Subspecies. Although satellite tracking suggests regional and not panmictic populations, there are no recognized subspecies.

Breeding biology. Loosely colonial with groups of up to 12 pairs. Nests are built high up in large trees in riparian areas, sometimes close to those of Hooded Vultures. In the Maasai Mara National Reserve in southern Kenya, nearest-neighbor nest distances averaged 553 yd and 1160 yd (511 m and 1071 m) in two separate areas. The species sometimes nests on power-line pylons. South of the equator, egg laying peaks in May at the onset of the dry season, and during October through December north of the equator. Pairs typically lay one-egg clutches, although two- and even three-egg clutches have been reported. Both parents care for eggs, which are incubated for an estimated 56 days. Nestlings fledge at 102–125 days, and remain partly dependent upon adults for an additional 5 to 6 months. Nesting success is estimated at 50–60%. In some areas as many as 20% of adults do not breed annually.

Diet and feeding behavior. The species specializes on meat and organ tissue when feeding gregariously on the carcasses of livestock, wild ungulates, and other large wildlife. As is true of other *Gyps*, individuals have serrated and grooved, shovel-shaped tongues that act together with cutting edges along the sides of it hooked bill, which allow individuals to quickly tear off and consume large, swallow-sized chunks of meat at rates of up to 1 lb/min (half a kg/min).

Social behavior. Less colonial during breeding than other African *Gyps*. Pairs presumably mate for life, with some reusing of individual tree nests for a decade or more. Post-fledging care appears to be limited to the current breeding season.

Movements. The movements of six immatures tagged at the Mankwe Wildlife Reserve in North West Province, South Africa, and tracked for at least 200 days, ranged across portions of South Africa, Botswana, Zimbabwe, and Zambia while establishing 100% MCPs of from 5400 mi^2 (14,000 km^2) to more than 228,000 mi^2 (588,000 km^2), and 50% KDEs of from 3400 mi^2 (8800 km^2) to 51,400 mi^2 (132,600 km^2). Individuals traveled an average of 126 mi (210 km) daily, with ranges shrinking during dry, winter months as ungulate mortality increased. A second tracking study in the Maasai Mara Natural Reserve, southern Kenya, demonstrated a link between foraging areas used and ungulate mortality during the dry season.

Conservation status. Populations of this once common and widespread, but now globally **Critically Endangered** species, face a number of threats, including both intentional and accidental poisoning, reduced carcass availability, and collisions with power lines and pylons.

Cape Vulture *Gyps coprotheres*

Spanish: Buitre de El Cabo
German: Kapgeier British
French: Vautour chassefiente

Taxonomy: Subfamily Aegypiinae

Size: Length 37–41 in (95–105 cm); wingspan 90–98 in (228–250 cm); 16–24 lb (7100–10,900 g)
Estimated world population: <10,000; declining
Movements: Adults largely sedentary; juveniles sometimes nomadic
Social ecology: Colonial nesting; communal nesting and roosting; group food finding

The Cape Vulture is the largest African *Gyps.* In part because of its large size, colonial nesting, and protracted population decline across many decades, the species has been the subject of more than 1500 scientific papers, popular articles, and reports and is the most studied African vulture. With a global population of fewer than 10,000 and a relatively small range of slightly more than 0.47 million mi^2 (1.2 million km^2), populations are currently confined to parts of Namibia, Botswana, Democratic Republic of the Congo, Angola, Zambia, Zimbabwe, Lesotho, Swaziland, Mozambique, and South Africa. Its protracted population decline, which may have slowed or even reversed recently in several parts of its range, is most likely attributable to several factors, including declines in the availability of large carrion, accidental poisoning, electrocution on power lines, and human take associated with witchcraft.

A massive vulture with relatively high wing loading, the species is a colonial cliff-nester whose numbers and breeding success have been monitored at a number of South African colony sites for decades. Although several nesting colonies are near-coastal, no records indicate that it feeds on marine-mammal carcasses.

The species was named *Gyps coprotheres* by the Scottish naturalist Johann R. Forster in 1798, based on a specimen collected in or near Cape Town, South Africa, by François Levaillant. The binomial was changed shortly thereafter to *Vultur kolbii,* in honor of the Dutch naturalist Peter Kolbe, and that name was used for more than a century until the earlier Forster description resurfaced, and the taxonomic principle of priority was invoked. Although the species differs anatomically from Indian Vultures and Slender-billed Vultures, molecular data suggest that it is part of a three-species clade with these species. The genus name *Gyps* is Greek for "vulture"; the species name *coprotheres* is Greek for "dung-hunting."

Subspecies. No subspecies are recognized, although it was once considered a subspecies of the Griffon Vulture.

Breeding biology. The species, which is generally monomorphic to slightly dimorphic, achieves adult plumage at about 6 years of age. Pairs, which appear to mate for life, often reuse the same nest site within colonies for many years, and some birds are colony philopatric. Females lay single-egg clutches. Eggs take 2 days to hatch after 57 days of incubation. Nestlings are guarded most of the time by at least one parent. Egg predators, the importance of which varies among colonies, include Verreaux's Eagles, White-neck Ravens, and baboons, with the latter sometimes destroying multiple eggs at a colony during a single visit.

Nestlings fledge at 124–171 days after hatching, and sometimes interact with adults post-fledging until the onset of the next breeding season. Confrontations can occur between fledglings and their parents, with adults jabbing at begging juveniles that have returned to the nest and preventing them from landing.

Although the species is believed to be an obligate cliff-nester, three of five adults satellite tracked in Namibia attended tree nests, including one individual that reportedly bred with a White-backed Vulture. Careful observations of 41 non-releasable and individually recognizable adults and subadults indicate that 22% of more than 300 observed copulation attempts involved extra-pair copulations, at least 23 of which resulted in cloacal contact, considerably more than suggested by earlier field observations.

Diet and feeding behavior. Prior to landscape changes that occurred after European colonization, the species is believed to have subsisted mainly on herds of migratory ungulates including springboks, elands, blesboks, red hartebeests, and both black and blue wildebeests, along with African elephants, giraffes, zebras, ostriches, and other large animals. Today carcasses of domestic livestock, including cattle, goats, horses, and pigs, are consumed in addition to wildlife carcasses. The species Latin name *coprotheres*, or dung hunter, probably results from the type specimen collected in Cape Town, which was most likely feeding on the carcasses of dogs and their dung at a garbage dump. Although Cape Vultures have been accused of taking live sheep and goats, it probably does so only on rare occasions when other nutritional resources are not available and is not likely representative of typical behavior.

Social behavior. Cape Vultures are colonial cliff-nesters whose colonies sometimes exceed 600 pairs, and whose breeding success often is linked to colony size. Field evidence suggests that breeding individuals feed mainly within 24–36 mi (40–60 km) of their colonies, typically in groups that number dozens of individuals, suggesting that colony-based social networks help individuals locate nutritional resources. Detailed observations indicate that nests surrounded by conspecifics produce more fledglings than those that are not, which supports that colonial cliff-nesting serves an anti-predator function, particularly in response to baboon predation. The extent to which colonial nesting helps enhance hunting skills in recently fledged and older immature birds is not known, but the aggressive behavior of adults toward the youngsters at the approach of the next breeding season, and the coincidental dispersive behavior of the young, suggest that parents do not assist in long-term development of hunting behavior. The degree to which such learning occurs among young in nursery areas, as well as the degree to which such areas facilitate pair formation, also remains unstudied.

Movements. The extent to which Cape Vultures migrate remains open to debate. Field evidence from the Eastern Cape Province of South Africa indicates high monthly counts of unaged individuals in December–May and then again in July–November, and a lack of sightings in April–June, suggesting migratory movements there. Re-sightings of individually marked immatures indicate they typically range farther than do breeding adults and that many subadults disperse into nursery areas away from active colonies, presumably to reduce competition with dominant adults that typically feed within 21–24 mi (35–40 km) of their colony sites.

Movements of satellite-tracked immatures indicate that dispersal movements increase slowly during the first 2 months after fledging and more rapidly thereafter, with a preference for areas with protected cliff faces. GPS satellite tracking of five adults in Namibia show home ranges with median minimal convex polygons (MCPs) of 8500 mi^2 (22,000 km^2), and that two juveniles ranged across areas averaging 187,000 mi^2 (482,000 km^2). Intriguingly, one of the juveniles exhibited a seasonal movement pattern with distinct dry- and wet-season ranges. Another tracking study involving five adult and four immatures in South Africa reported adult MCP home ranges of 47,000 mi^2 (122,000 km^2) and immature home ranges of 191,000 mi^2 (492,000 km^2), confirming substantial age-related differences in movements.

Conservation status. Several populations of Cape Vulture have been in decline for more than 40 years, and coincidental range retractions have been known to occur. Although threats, including electrocution along power lines and both targeted and accidental poisoning, have increased recently, the long-standing effect of reduced carcass availability in this largely central-place foraging, colony-nesting species remains an intrinsic concern. That numbers breeding at several traditional large South African colonies have remained steady or increased recently suggests that at least some populations may be increasing in response to increased food availability. First listed as **Threatened** by BirdLife International in 1988, the species is now considered **Endangered** globally.

Red-headed Vulture *Sarcogyps calvus*

Spanish: Buitre Cabecirrojo
German: Kahlkopfgeier
French: Vautour royal
aka: King Vulture, Pondicherry Vulture

Taxonomy: Subfamily Aegypiinae

Size: Length 30–34 in (76–86 cm); wingspan 79–89 in (200–225 cm); 8.0–12.0 lb (3700–5400 g)
Estimated world population: <15,000; declining in parts of range
Movements: Largely sedentary, but sometimes wide-ranging
Social ecology: Largely solitary; territorial nesting and roosting; group food finding

One of the least studied of all Old World species, the largely solitary Red-headed Vulture has a species range of slightly more than 2 million mi^2 (5 million km^2). A medium-sized vulture with blackish plumage; a large, triangle-shaped, orangish-to-reddish featherless head; and reddish legs, in which adult males have pale yellow irises and adult females have dark brown irises. The species was once relatively common in the western foothills of the Himalayas, and less so in Southeast Asia as far south as Singapore. Described as timid compared with *Gyps*, Red-headed Vultures are typically seen alone or in groups of several pairs. A precipitous population decline in the 1900s into the 2000s, presumably caused by diclofenac poisoning, followed shortly after that of the southern Asian *Gyps* species.

The species was named *Sarcogyps calvus* by the Austrian naturalist Giovanni Antonio Scopoli in 1786, based on a specimen collected in Pondicherry, India. Although sometimes included in the genus *Aegypius* or *Torgos*, the Red-headed Vulture differs sufficiently to merit placement in its own monospecific genus. The genus name *Sarcogyps* is Greek for "flesh-vulture"; the species name *calvus* is Latin for "bald."

Subspecies. No recognized subspecies.

Breeding biology. Territorial, with aerial displays including mutual soaring and talon grappling. Builds stick nests high in trees, which it often decorates with rags, pieces of animal skin, and wool. Almost always single-egg clutches, incubated for 58 days. Both sexes share incubation and brooding, and both provide food to developing young. Nestlings fledge at about 4.5 months. Post-fledging parental care and fledgling behavior are unstudied.

Diet and feeding behavior. Mainly carrion but also feeds on injured birds, fish in desiccating puddles, and on small-animal victims of grass fires. Its occurrence is said to be correlated with that of mammalian predators, including tigers. The species dominates Egyptian Vultures and small, but not large, groups of *Gyps* vultures at carcasses.

Social behavior. Largely territorial. No reports of colonial nesting or roosting.

Movements. Territorial and sedentary. Young may be nomadic.

Conservation status. Given the species' current rarity in Southeast Asia, where only a few hundred likely remain, together with recent declines in India caused by diclofenac poisoning, BirdLife International lists the Red-headed Vulture **Critically Endangered** globally.

White-headed Vulture *Aegypius occipitalis*

Spanish: Buitre Cabeciblanco
German: Wollkopfgeier
French: Vautour à tête blanche

Taxonomy: Subfamily Aegypiinae

Size: Length 28–32 in (72–82 cm); wingspan 80–89 in (205–225 cm); 7.0–12.0 lb (3300–5300 g)

Estimated world population: <6,000; decreasing
Movements: Sedentary; immatures more nomadic
Social ecology: Solitary, territorial nesting; individual and group food finding

A medium-sized, visually distinctive, and sexually dimorphic African endemic that occurs largely in protected areas within its 8 million mi^2 (21 million km^2) range. With a distribution said to be linked to that of the baobab tree, the White-headed Vulture inhabits mixed deciduous woodlands. Until recently, little studied, in part because of its relative scarcity and solitary nature.

Juveniles are mainly dark brown. Adults are largely blackish above, with conspicuous buff-edged, medial wing coverts; white belly and femoral tract; a white, downy head; red beak; blue cere; and a bare, pinkish face. Adult females are slightly more massive than males and have white inner secondaries that are gray in males. Adults are most frequently seen alone or in pairs. Younger birds sometimes congregate in nursery areas.

The species was named *Vulture occipitalis* by the English naturalist William Burchell in1824, based on a specimen collected in the Northern Cape Province of South Africa. The genus name was changed to *Trigonoceps*, after its somewhat triangle-shaped skull, by French Naval Surgeon René Primevère Lesson in 1842, and to *Aegypius* in 1951 as suggested by C. M. N. White, due to similarity to other members of that genus. The genus name *Aegypius* is Greek for "vulture"; the species name *occipitalis* is Latin for "back of the head."

Subspecies. No recognized subspecies.

Breeding biology. The White-headed Vulture is an often territorial, sedentary species, whose pair bonds and nest sites sometimes extend across years. Pairs build and refurbish bulky stick nests at the top of trees or in the uppermost fork. One-egg clutches are incubated by both parents for 55–56 days. Nestlings are fed by parents with beak-to-beak via regurgitation and by parents depositing food on the floor of the nest. Nestlings, which fledge at about 120 days, are fed by their parents for up to 5 months.

Australian raptor specialist Campbell Murn and colleagues studied breeding ecology across more than 10 years in Kruger National Park, where laying occurred earlier in southern versus northern areas of the park in June and July. Pairs were territorial and nearest-neighbor distances averaged approximately 0.6 mi (1 km). Breeding success was 0.69 chicks per active nest.

Diet and feeding behavior. The species feeds on both small and large carcasses, typically at the perimeters of groups of *Gyps* vultures. Food remains at nests, including steenboks, striped weasels, hares, and tortoises, together with observations of birds feeding on the eggs, nestlings, and adults of Greater Flamingoes and Lesser Flamingoes, have long suggested that the species is sometimes predatory

The most detailed report of predatory behavior includes observations by Campbell Murn, who watched an adult pair take a living slender mongoose, an adult female and juvenile pursue and catch a monitor lizard, and an adult male and a pair of adults chase tree squirrels. Murn's observations suggest that the individuals involved were well practiced in predatory pursuit, and that cooperative hunting may be common in the species. An observation of two individuals catching a young impala in northwestern Zimbabwe suggests that cooperative behavior may be widespread. The species' visual field more closely resembles that of predatory raptors than that of *Gyps* vultures in that they have a wider binocular field that most likely enhances accurate location of its talons when seizing live prey.

Social behavior. Aside from the observations above, the extent to which the species is social remains unstudied.

Movements. Seasonal movements of juveniles in Chad have been suggested based on the fact that ratios of juveniles-to-adults there increase to about 10:1 during the rainy season.

Conservation status. Recent studies suggest that the estimated global population of <6000 individuals is declining rapidly, mainly due to poisoning, human persecution, and ecosystem changes. BirdLife International lists the White-headed Vulture as **Critically Endangered** globally.

Cinereous Vulture *Aegypius monachus*

Spanish: Buitre Negro
German: Mönchsgeier
French: Vautour moine
aka: Black Vulture, Monk Vulture

Taxonomy: Subfamily Aegypiinae

Size: Length 39–47 in (100–120 cm); wingspan 98–116 in (250–295 cm); 15–25 lb (7000–11,500 g)
Estimated world population: <32,000; decreasing
Movements: Partial migrant
Social ecology: Solitary nesting; communal roosting; individual and group food finding

A massive, large-beaked, broad-winged, dark brown vulture with a pale head. Juveniles are darker, overall. One of the largest Old World vultures, the species has a global range of more than 8.5 million mi^2 (22 million km^2) that stretches from Iberia and Morocco, east across southern Europe, northeastern-most Africa, into the Middle East and across central Asia into northern China, Mongolia, and Manchuria. Cinereous Vultures inhabit mainly hilly and mountainous areas where it feeds almost exclusively on carrion in high-altitude meadows and other open areas, but it sometimes takes living prey, including rabbits and other small mammals. Unusual among vultures, females are slightly larger than males. The species is a partial migrant at higher latitudes and high-mountain regions.

The Cinereous Vulture was named *Vultur monachus* by the Swedish taxonomist Carolus Linnaeus in 1766, based on a specimen collected in Arabia described by the English naturalist George Edwards. The genus name was changed to *Aegypius* in 1809 by French zoologist Marie Jules-César Savigny. The genus name *Aegypius* is Greek for "vulture"; the species name *monachus* is Latin for "hooded."

Subspecies. No subspecies are recognized; however, individuals in China and Mongolia typically measure larger than elsewhere.

Breeding biology. Cinereous Vultures build enormous flat-topped nests that are refurbished annually at the tops of oaks, almonds, junipers, and other conifers and on cliffs in treeless areas. Males and females share chick-rearing duties. Single-egg clutches are the norm, but two egg-clutches also occur. Incubation is estimated at 55 days. Fledging occurs at from 105 to 120 days post-hatching. Parental care extends for 2 to 3 months post-fledging. In Mongolia, predators at nests include Pallas Cats, Northern Ravens, Spanish Imperial Eagles, and Bearded Vultures. Fledgling dispersal occurs.

Diet and feeding behavior. Diet varies considerably, both geographically and seasonally. The species routinely feeds on the carcasses of large domestic and wild ungulates, sometimes with *Gyps* species, which it dominates when not overwhelmed numerically. It also takes living prey, including rodents, lagomorphs, tortoises, terrapins, and other reptiles.

Analysis of regurgitation pellets reveal both spatial and temporal differences in diet. Regurgitation pellets collected in Turkey in 2004 indicate sheep in 77% of the specimens, together with human-hunted remains of wild boar in 44%, and domestic chicken in 23%, with the latter increasing near poultry farms. Regurgitation pellets collected in 2000 in Extremadura, Spain, indicate sheep in 58%, domestic and wild boar in 16%, and hunted deer in 14%, with the latter increasing during deer hunting. Rabbits were found in 78% of regurgitation pellets collected in Extremadura in 1970 when viral myxomatosis was decimating rabbit populations there.

Social behavior. Little studied. Territorial when breeding. Limited communal roosting.

Movements. Cinereous Vultures are partial migrants in Mongolia, Iberia, and the Caucasus range. Like Rüppell's Vultures, the species has high-altitude hemoglobin that functions effectively at low partial pressures of oxygen encountered at high altitudes, along with one that functions at lower altitudes.

Juvenile dispersal appears significant. Satellite tracking and wing tagging indicate that juveniles and subadults from Mongolia overwinter on the Korean peninsula, where feeding stations routinely attract dozens of individuals. Juveniles from the Caucasus range overwinter as far south as northeastern Saudi Arabia and southwestern Iran. Those from central Turkey migrate south as far as the Arabian Peninsula. Individuals from central Asia overwinter in central and southern India. In western Europe, migrants have been reported crossing the Strait of Gibraltar, and at least one immature is known to have overwintered in Senegal. Twelve satellite-tracked juveniles from central Spain maintained home ranges of 31,000 mi^2 (81,000 km^2) that averaged 104 mi (174 km) from their nests. A dispersing wing-tagged juvenile from East Gobi, Mongolia, traveled more than 1620 mi (2700 km) north-northeast to Yakutsk, Russia, at more than 62° N latitude, 3000 mi (5000 km) east of Moscow. The species is a winter vagrant in Myanmar, Thailand, Cambodia, peninsular Malaysia, and the northernmost Philippines.

Conservation status. Numbers are declining in several Asian populations but are increasing in Europe. BirdLife International lists the species as **Near Threatened** globally.

Lappet-faced Vulture *Aegypius tracheliotos*

Spanish: Buitre Orejudo
German: Ohrengeier
French: Vautour oricou
aka: King Vulture, Nubian Vulture, Eared Vulture, Sociable Vulture

Taxonomy: Subfamily Aegypiinae

Size: Length 37–45 in (95–115 cm); wingspan 98–114 in (250–290 cm); 12–21 lb (5400–9400 g)
Estimated world population: <10,000; declining
Movements: Largely sedentary; juveniles and subadults nomadic
Social ecology: Relatively solitary, territorial breeding; communal roosting; individual and group food finding

A massive, large-billed vulture that, except for scattered chin "whiskers," features a pinkish featherless head that blushes reddish with excitement, and prominent lateral, wattle-like folds. Adult Lappet-faced Vultures are largely blackish-brown above, and heavily streaked blackish below, except for white leggings in the nominate race. Juveniles are browner and plainer overall.

With a global population of <10,000, and a 12 million mi^2 (32 million km^2) largely sub-Saharan range that includes the Senegal, the Sahel, and parts of Sudan, Egypt, and the Arabian Peninsula, the species is not especially common anywhere. The most spontaneously aggressive of all African vultures, the species routinely dominates other species at carcasses, where pairs often fight with conspecifics.

Territorial pairs breed in semi-arid and mesic open savanna and along mountain slopes where junipers and other flat-topped trees provide stable platforms for the species' enormous multiple-year stick nests. The species also breeds in crags in treeless areas. Lappet-faced Vultures feed almost exclusively on carrion but also take flamingo eggs, as well as termite alates and locusts. At least in the northern *negevensis* subspecies, females are slightly larger than males.

The type specimen was collected by the French explorer and ornithologist Francois Levaillent in southern Namibia and named *Vultur tracheliotos* in 1791 by the German explorer and scientist Johann Reinhold

Forster. The genus name was changed to *Torgos* in 1809 by the German naturalist John Jakob Kaup in 1828, and to *Aegypius* in 1951 as initially suggested by C. M. N. White, and later supported by the American taxonomist Dean Amadon and the African raptor specialist Leslie Brown because of similarity to other members in that genus. The genus name is Greek for "vulture"; the species name *tracheliotos*, which is sometimes spelled *tracheliotus*, is Greek for "eared throat," in reference to its prominent wattles.

Subspecies. Two subspecies are recognized. The nominate *tracheliotos* is blackish overall, with prominent lappets, white leggings, and a black bill; the more northern *negevensis* is browner overall, with less-prominent wattles, partly brown leggings, and a blackish bill. West African individuals resemble the *tracheliotos*. Both subspecies are similar in appearance as immatures, and the two are sometimes considered extremes of a latitudinal cline.

Breeding biology. Like Cinereous Vultures, pairs build and annually refurbish enormous flat-topped nests, typically at the tops of short acacias, and in crags. Nearest-nest distances average 2.3 mi (3.9 km) in the Arabian Peninsula and 3.9 mi (6.5 km) in Swaziland. Both males and females share incubation and chick-rearing duties, including shading on sunny days. Single-egg clutches are the norm, but two-egg clutches also occur. Incubation is estimated at 55–56 days. Fledging occurs at 105–120 days post-hatching, with parental care extending for up to 3 additional months. Nesting success of 52 monitored pairs in Saudi Arabia was 56%. Nest predators are unstudied.

Diet and feeding behavior. The species routinely participates in multi-species feeding groups at carcasses. Although it is sometimes the first to open large carcasses with its massive bill, it often arrives at carcasses after *Gyps* have begun feeding, and it remains on the perimeter of feeding groups. It approaches the carcass after other vultures have finished feeding, at which time it consumes skin, tendons, and ligaments, as well as the skull and its contents, which it crushes with its oversized beak. The remains at nests suggest predation on a variety of small animals, including monitor lizards, pangolins, steenbok, duiker, goats, polecats, jackals, civets, and mongooses. The species also preys on the eggs, nestlings, and fledglings of flamingoes, taking the latter, in flight, cooperatively with other Lappet-faced Vultures. The species routinely visits, bathes, and drinks at watering holes.

Adult Black Vulture. This species is the most social of all New World Vultures.
(Florida; photo by Shawn Carey)

Adult Black Vulture (*left*) and adult Turkey Vulture (*right*). These two species currently have the largest contiguous ranges and are the most abundant of all vultures.
(Florida; photo by Shawn Carey)

Subadult Turkey Vulture. Note the partial red head and the dark-tipped bill.
(Florida; photo by Shawn Carey)

Lesser Yellow-headed Vulture. Like its close relative the Turkey Vulture, this species uses smell as well as sight to locate carrion. (Argentina; photo by Sergio Seipke)

California Condor. Adult and juvenile. This species is the most critically endangered of all vultures. (California; photo by Mike Lanzone)

Juvenile California Condor. Note the dark head and downy feathers above the nostrils. (California; photo by Michael Lanzone)

Adult male Andean Condor. Note the fleshy comb and full crop.
(Argentina; photo by Ignazio Gonzalo)

Adult female Andean Condor. Note the lack of a fleshy comb.
(Argentina; photo by Ignazio Gonzalo)

Adult Palm-nut Vulture. This species is the most tropical of all African Vultures.
(South Africa; photo by Andre Botha)

Adult Bearded Vulture. Note the bone fragment in its talon.
(South Africa; photo by Shane Elliot)

Egyptian Vulture. The bright yellow facial skin of this species is derived from carotenoids ingested from the fecal material of cattle.
(Ethiopia; photo by Michael J. McGrady)

Egyptian Vulture. This species opens ostrich eggs by cracking them with thrown stones.
(Ethiopia; photo by Michael J. McGrady)

Adult Hooded Vulture. One of the smallest African Vultures, this species is a relatively abundant human commensal in West Africa.
(South Africa; photo by Andre Botha)

Adult Indian Vulture. This once-common species is now Critically Endangered due to poisoning by the veterinary drug diclofenac.
(India; photo by Naveen Pandey)

Adult Slender-billed Vultures (*upper right* with wings spread) and adult White-rumped Vultures (*left*).
(Nepal; photo by Tulsi Subedi)

Adult Griffon Vulture, the most widespread and abundant of all vultures in Europe. (Spain; Fundacion Migres archives)

Cape Vulture. This African endemic breeds in southernmost Africa. (South African photo by Andre Botha)

Adult Himalayan Vulture. This species is the largest Asian *Gyps*. (Nepal; photo by Tulsi Subedi)

Adult Cape Vulture. This African endemic species nests colonially in cliffside colonies.
(South Africa; photo by Andre Botha)

Adult Red-headed Vulture. This species is one of the least studied of all Old World vultures.
(India; photo by Munir Virani)

Adult, female White-headed Vulture. This species is largely territorial and sometimes predatory.
(South Africa; photo by Andre Botha)

Nestling Cinereous Vulture with wing tag.
(Mongolia; photo by *Richard Reading*)

Adult Cinereous Vulture. Although widespread in Europe and Asia, this partial migrant is relatively uncommon. (Mongolia; photo by Richard Reading)

Social behavior. Although once referred to as the "Sociable Vulture," in part because of an early description of individuals following one another to carrion, the species is relatively solitary, territorial, and routinely expresses dominance over conspecifics and other species at carcasses. Pairs, which often fly together when searching for food, typically perch together at carcasses and at watering holes. Pairs defend nests together year-round and fight with conspecifics at watering holes

Movements. Some individuals engage in seasonal movements; however, adults are often sedentary. In Saudi Arabia, two satellite-tracked subadults left their nesting area for about 7 months post-tagging before returning, after which one continued to move annually, whereas the other was largely sedentary except for a single long-distance movement of more than 360 mi (600 km). Six subadults satellite tagged in the Maasai Mara National Reserve in Kenya and followed for an average of 146 days ranged over smaller areas than did tagged African White-backed Vultures and Rüppell's Vultures, and they were less likely to follow the regional movements of wildebeest. By far, the most extensive study of satellite-tagged birds involved 14 adults in Botswana and was undertaken by Rebecca Garbett, who found that during the breeding season, home ranges were eight times smaller in breeding versus non-breeding individuals, with 95% home-range KDE estimates of 6600 mi^2 (17,000 km^2) versus (54,000mi^2) (139,000 km^2), respectively. In addition, annual 95% KDEs and MCPs for all birds, and averaged 76,000 mi^2 (195,000 km^2) and 97,000 mi^2 (250,000 km^2), respectively, and breeders selected protected areas year-round, whereas non-breeders selected protected areas only during the breeding season. Garbett considered the individuals she studied, many of which spent nearly half their time outside of Botswana, to be partial migrants.

Conservation status. With a small and rapidly declining global population of <10,000, and with declines attributed to poisoning, persecution, and ecosystem change, BirdLife International lists the species as **Endangered** globally.

3 PAIR FORMATION AND REPRODUCTION

> Reproduction is an essential part of raptor biology.
> Without reproduction, there will be no next generation.
> And without a next generation, extinction will follow.
> **Ian Newton (1986)**

ALTHOUGH WE KNOW more about the breeding ecology of many predatory raptors than about any other aspect of their ecology, the same is not true of scavenging raptors, for which feeding behavior is far better understood than breeding behavior. And indeed, detailed nesting studies of many species, including King Vultures, Greater and Lesser Yellow-headed Vultures, Slender-billed Vultures, Indian Vultures, White-rumped Vultures, and Red-headed Vultures, have yet to be detailed. Nevertheless, much of what we do know about the breeding biology of vultures makes sense in light of their sizes, life spans, and deferred maturations.

As is true of most birds, obligate avian scavengers time their breeding activities to take advantage of seasonal fluctuations in food availability, which—excepting Palm-nut Vultures, whose breeding ecology is largely tied to the seasonal availability of palm nuts—tend to be linked to the twin availability of carcasses and atmospheric updrafts necessary for the soaring flight used in food-searching. Most vultures take these availabilities into account as they court, pair, build nests or prepare nest scrapes, lay and incubate eggs, brood, feed and protect young, and then protect and feed them for at least several weeks after they fledge. That said, notable differences occur between obligate scavenging and predatory raptors, as well as notable differences among species. Much remains to be learned about this aspect of vulture ecology, and what follows is largely indicative, rather than encyclopedic.

SIZE DIMORPHISM

One obvious anatomical difference between predatory raptors and scavenging birds of prey is that whereas almost all species of predatory birds of prey have females that are notably larger than their male counterparts, the masses of female and male vultures are relatively similar, with four notable exceptions. Adult females are measurably larger than males in both White-headed Vultures and Lappet-faced Vultures, two species of occasionally predatory and territorial vultures, as well as in frequently siblicidal Bearded Vultures. Male Andean Condors, however, are substantially more massive than their female counterparts.

The situation in which females are larger than their male counterparts—an unusual trait in birds—is largely limited to species of particularly predatory gulls and seabirds including skuas, frigate birds, and boobies, as well as in most all diurnal birds of prey and many owls. Female hummingbirds, too, are larger than their mates, probably because of size considerations associated with minimum viable egg size.

In predatory raptors, differences in body mass range from 15% to 60%. Because males are larger than females in most mammals, including humans as well as most birds, biologists call this unusual gender-related difference in size reversed size dimorphism (RSD).

Raptor biologists have tried to explain RSD in many ways. Most modern hypotheses fall into one of three categories: ecological, behavioral, or physiological-anatomical. Although dozens of explanations have been proposed, none has gained universal acceptance. One reason for the lack of consensus is that although many current hypotheses explain why male and female raptors should differ in size, most fail to explain why it is the female, rather than the male, that is larger, and why it occurs only in raptors and a few other groups of birds.

Most of the ecological hypotheses depend on factors that affect adult individuals. One popular explanation, for example, suggests that RSD acts to reduce competition between members of breeding pairs by allowing larger females to take larger prey and smaller males to take smaller prey. And indeed, field evidence indicates that females often do take larger prey as compared with the males. But this explanation fails to explain why it is the female, and not the male, that is larger in raptors. Another explanation focused on differences in adults is that smaller males are more maneuverable than the larger females and, thus, are better able to catch smaller and usually more numerous prey. This ability might be important during the breeding season when males do most of the hunting for their mates and young. Presumably, females are larger than males, either to better incubate their eggs and brood their young, or to reduce competition with males outside of the breeding season. But the question becomes: Why isn't RSD the rule in most other birds?

An additional hypothesis that focuses on selection pressure on nestlings, rather than adults, merits note.

The "head start" hypothesis is based on the following lines of reasoning: 1) Male raptors, which typically provide the overwhelming majority of prey during the breeding season, must be particularly proficient hunters if they are to breed successfully. Field evidence suggests that males do hunt more successfully than females. 2) Both females and males are under intense selection pressure to breed as early in life as possible, as this increases their lifetime reproductive success of overall fitness. This pressure means that as the principal food provider during the breeding season, males will need to develop hunting skills as quickly as possible. And, again, field evidence indicates that in most raptors, males do grow up faster than their larger female siblings; and furthermore, they leave the nest earlier than females and begin learning how to hunt earlier than females. 3) If males were larger than nestling females, or even if they were the same size as females, selection for more rapid development in the first year of life could place their female siblings at a disproportionate risk of being killed by nestling males, a phenomenon that biologists call siblicide. 4) By being smaller than their female siblings, males reduce this last risk while at the same time enhancing their own rapid development. That other species that exhibit RSD also exhibit siblicide supports the head-start hypothesis, which makes a series of additional predictions that have yet to be tested. Nonetheless, this final line of reasoning remains a leading candidate for explaining RSD in predatory birds of prey.

On the one hand, given that two of the largest and more solitary obligate scavengers, White-headed Vultures and Lappet-faced Vultures, exhibit RSD, as does the sometimes siblicidal Bearded Vulture, suggests that one of the explanations offered above may be responsible for the phenomenon in vultures and condors. On the other hand, that all three of these species are Old World members of the avian family that includes the kites, harrier, hawks, and eagles of the world, suggests that phylogenetic inertia also may be involved. An evolutionary phenomenon first described by Charles Darwin, and afterward named as such by B. Huber in 1939, phylogenetic inertia is said to constrain future evolutionary pathways, and it too may play a role in RSD. Darwin suggested that the organisms do not start as blank slates, but rather with characteristics that build upon existing ones inherited from their ancestors; and that these characteristics likely limit evolution overall. Whatever the explanation for its spotty occurrence in obligate scavenging raptors, RSD merits additional study.

BREEDING BIOLOGY

Deferred Maturity

As in many large-bodied predatory birds of prey, vultures exhibit deferred or delayed maturation, a behavior in which individuals breed only after reaching full adult plumage at 4 to 6 years of age. Some researchers have suggested this trait is attributable to the likelihood of it taking several years to develop the carcass-securing and, in some instances, territorial skills needed for successful breeding. Captive-bred females tend to breed at earlier ages than males, which is a behavior typical of species with deferred maturation and that may be attributable to the fact that the females' role in food-finding, territorial defense, or both, is less than that of males; or that male mortality is higher than female mortality.

Although information on the age of first breeding in the wild is limited in vultures, in well-studied species, information appears to be considerable. Black Vultures and Turkey Vultures apparently first breed at 8 and 6 years, respectively; and initial breeding in a large population of captive Bearded Vultures averaged 7.7 years in females and 9.7 years in males. Exceptions include non-migratory island populations of Egyptian Vultures, which sometimes breed in subadult plumage, most likely because the small size of island populations favors earlier breeding because of the inherent risk of extirpation in this small, insular population.

Courtship

Given the high occurrence of long-term, multiple-year pairings, annual courtship is not particularly elaborate in most scavenging raptors. It does, however, occur. In both Black Vultures and Turkey Vultures, for example, pairs routinely sit together conspicuously on trees or buildings near nest sites for several weeks while engaging in ritualized tandem territorial aerial display flights, with the male following 65–165 ft (20–50 m) behind and typically above the female. Flights are interspersed with on-the-ground, wing-spread dancing and bill-clicking followed by copulation and mutual preening. Tandem flight also occurs in California Condors and Griffon Vultures, as does aerial talon-grappling and cartwheeling in which individuals lock talons and spin together while losing altitude. Although the latter is most often associated with aggressive territorial behavior, it also apparently functions in courtship, at least in Bearded Vultures, Palm-nut Vultures, Hooded Vultures, and Cinereous Vultures.

In both White-backed Vultures and Cape Vultures, males frequently utter "distinctive hoarse calls" when holding on and balancing on the backs of females when copulating. These two species, as well as Palm-nut Vultures, Egyptian Vultures, Bearded Vultures, Indian white-rumped Vultures, Rüppell's Vultures, and Griffon Vultures, engage in dozens, if not hundreds, of copulations in the weeks leading up to egg laying. Some speculate that frequent copulations function in mate familiarity or in territorial signaling, whereas others have suggested that it might serve to reduce the potential for extra-pair copulations (EPCs) in which non-paired males and females copulate with each other when the female is mated with another male. Although EPCs are thought to be rare in most vultures, they do occur in introduced populations of California Condors, for which 23% of all copulations in two breeding seasons between five males and seven females involved birds that were not "social mates." Females often rebuff EPCs, but some EPCs do result in inseminations.

Although males typically benefit genetically from such events, genetic benefits to females are less clear. This uncertainty is particularly so in species that lay one-egg clutches, when a single offspring might be fathered by a male that made no contribution to its successful rearing. Some researchers argue that males engaging in extra-pair copulations may offer higher genetic quality sperm than do other males; however, the extent to which this occurs is unclear. In any case, the possibility of EPCs may explain why mated pairs engage in high numbers of copulations in the weeks leading up to egg laying.

Pair Fidelity

Multi-year pair fidelity, a phenomenon thought to be common in vultures, has been little studied because of limited observations of individually marked birds. Although Turkey Vultures were once believed to mate for life, recent studies involving camera traps and marked individuals in a population nesting in abandoned buildings in Saskatchewan, Canada, indicate mate switching occurred with a female visiting and presumably breeding in three different buildings in single years; with another marked female visiting or residing in three different buildings and copulating on the roof with a fourth male, while raising young at different nest sites within several kilometers with at least two different males; and with another marked pair ejecting a prospecting pair from a site that the former had previously used. Taken as a whole these observations suggest both flexibility and continuity, in both pair and nest-site fidelity. By comparison, individually marked Griffon Vultures in a reintroduced population in France exhibited strong nest-site and mate fidelity, with 95% of 22 pairs of marked individuals remaining together and at the same site across multiple years. Overall, pair fidelity in vultures seems likely in most species where nesting sites are limited.

Territoriality

Introduced to ornithology in 1920 by the amateur English ornithologist H. Eliot Howard, *territoriality*, or territorial behavior, describes an area of exclusive intraspecies use that pairs of individuals defend from others of their species. Presumably this occurs because doing so provides adequate nutritional resources, helps protects the nest from predators, and reduces the likelihood of EPCs. Territories occur for most birds, as well as for many vertebrates, invertebrates, and organisms as simple as gelatinous masses of nucleated protoplasmic slime mold. That said, territorial behavior, which remains little studied in many vultures, apparently varies considerably within and among species, depending on both the quality of the site in question and its defensibility. High-quality nesting areas are in particular demand. As in most birds, temporal priority of territorial establishment provides an advantage in displacement attempts. Colonial-nesting species, including Griffon Vultures and Cape Vultures, defend only the nest itself and the area within several meters of it, whereas African White-backed Vultures sometimes defend the entire nesting tree. Black Vultures tolerate conspecifics near nests more so than do Turkey Vultures, with as many as 20 pairs nesting on a 1.4-ac (0.6 ha) island in South Carolina, and individuals nesting as close as 75 yd (70 m) in southern Pennsylvania. Territorial behavior has been best studied in Egyptian Vultures and Bearded Vultures, with territoriality in the latter being linked to the fact that this species sometimes hoards uneaten bones in ossuaries close to their nests and guards them from theft by Griffon Vultures. Although Lappet-faced Vultures are often considered wide-ranging, movements of seven breeding and seven non-breeding birds differed during the breeding and non-breeding season, with breeders maintaining 95% kernel density estimate (KDE) home ranges eight to ten times smaller than those of non-breeders. My observations in Kenya, where individuals apparently refurbish and reuse huge stick nests across years, indicate that pairs typically arrive at carcasses in series and often engage in vicious and prolonged between-pair fights involving targeted biting, foot-stabbing, and body tossing, which greatly exceed the intensity of those among *Gyps*. None of my observations involved marked individuals; however, it was my impression that the intensity and persistence of such encounters was territorial and not just squabbling over a meal.

Breeding Success

Breeding success, or the number of young produced by each breeding pair annually, is particularly difficult to assess, in part because early-season failures are often undetected. Field evidence suggests that factors including food availability, weather, and predation—acting alone or together—influence the likelihood of successful breeding. Unfortunately, most fieldwork has focused on assessing breeding success in areas with relatively dense and relatively successful breeding populations that are likely to represent maximal or near maximal estimates, rather than typical estimates, of reproductive success. Taking that into account, a summary of breeding success in 10 species of African species indicated that four species raised at least one nestling at least 40% of the time, and that the remaining six species did so 50 to 91% of the time. Similar data for non-African species are not as available, but an analysis of nesting records for two North

American species indicates overall breeding success of 53% for Turkey Vultures and 37% for Black Vultures.

SYNTHESIS AND CONCLUSIONS

1. Reproduction is an essential aspect of the biology of scavenging birds of prey. Without it there would be no next generation, and extinction would follow.
2. Vultures breed seasonally, especially in response to carcass and updraft availability.
3. Excepting several larger and often more predatory species in which females are notably more massive than their male counterparts, males and females differ little in body mass, except for Andean Condors, in which males are larger than females. The Bearded Vulture, a species in which siblicide occurs, also exhibits reversed sexual size dimorphism, that is, the females are larger.
4. Almost all vultures and condors delay maturity, most likely because it takes multiple years to develop carcass-searching and territorial skills necessary for successful breeding.
5. Multi-year pair fidelity is typical in scavenging raptors. Extra-pair copulations do occur, as does within-year mate switching.
6. Territoriality occurs in some, but not all, vultures.
7. Breeding success, which varies with food availability, weather, and predation, appears to be high in many species, although this may be so because it has been studied in areas with relatively dense and, presumably highly successful, populations.

4 FOOD FINDING AND FEEDING BEHAVIOR

> Old World Vultures depend upon a food resource that is scarce and unpredictable in space and time. This dependency has led to a foraging strategy based on minimizing the energetic costs of searching, and maximizing the range of foraging flights.
>
> **F. Hiraldo and J. A. Donázar (1990)**

AS IN MOST things ecological, feeding on carrion has both benefits and costs. Although many birds do it, most are facultative scavengers, meaning largely predatory species that feed on carrion as a backup. Only a few dozen avian scavengers depend on carrion feeding either exclusively, or nearly exclusively, for their food. Obligate avian scavengers include the 23 species of Old and New World vultures, birds whose diets are believed by many to be both spatially and temporally unpredictable. Although the scarce and unpredictable nature of carrion has been overstated by many scientists, carrion feeding has favored the development of low-cost searching flight in vultures.

THE CARRION RESOURCE

Carrion, animal dung, and nutritional items in human trash provide several benefits as food, including digestibility and nutritional quality. Unfortunately, they also provide several costs, including an increased likelihood of infectious diseases and harmful toxins. Vultures have evolved a number of anatomical, behavioral, and physiological adaptations to deal with and to take advantage of the distribution and abundance of carrion and other organic waste as nutritional resources in ways that allow scavenging raptors to exist in all but high-latitude terrestrial ecosystems.

Carrion Availability

At first blush, few would argue with the perceived scarce and unpredictable nature of carrion. After all, when walking or driving through pastureland crowded with livestock, one rarely sees dead individuals; excluding, of course, road kills, and most nature documentaries depict carrion only at the conclusion of successful predation events. Carrion, it would seem, is a genuinely rare and unpredictable commodity. But death is a universal part of life, and in most instances the rate at which it occurs is predicable, both in time and in space across most landscapes, natural and human-dominated. Sooner or later, all animals die, and when they do, carrion becomes available to vultures, condors, and other scavengers. And in one way or another, scavengers manage to find it.

Box 4.1 Key Innovations, Niche Expansions, and Interspecies Competition between *Cathartes* and Black Vultures

Three species of *Cathartes* vultures, the Greater Yellow-headed Vulture, the Lesser Yellow-headed Vulture, and the Turkey Vulture, possess two exceptional adaptations that markedly increase their food-finding skills. The first is their well-known sense of olfaction. The second, less discussed and less appreciated, is their low-altitude, contorted soaring behavior. Together these abilities function as "key innovations," a term introduced in 1949 by the American evolutionary biologist Alden Miller to describe the development of evolutionary traits that foster both species diversity and ecological success. Prime examples of key innovations include the development of flapping or "powered" flight in bats and buoyancy-enabling swim bladders in bony fishes, both of which acted to substantially increase species diversity and ecological success in those taxa.

The most widely recognized and appreciated key innovation in *Cathartes* vultures, the only multi-species genus of vultures in the New World, is a well-developed sense of smell, a vulturine trait that allows three *Cathartes* species to seek out and pinpoint carcasses by smell as well as by sight. A second and less widely acknowledged key innovation, the dihedral posture of their upswept wings, allows them to effectively extract atmospheric energy in boundary-layer turbulence.

Like harriers and *Milvus* kites, both of which search for food in low altitude, boundary layer, non-flapping, rocking flight, *Cathartes* vultures normally hold their wings above their backs in a dihedral posture while soaring. The Egyptian Vulture, another often low-flying vulture also flies in a dihedral, although far-less pronounced. All of these species routinely search for living prey or carrion while flying to within several meters of the vegetative canopy, airspace with considerable small-scale, near-surface, boundary layer turbulence. In such situations, soaring and gliding in a dihedral is more efficient than flight on horizontal wings.

Consider what happens when one wing in the dihedral receives a small upward puff of turbulent air and is pushed upward. This repositioning of the flight profile immediately reduces the lift of the upswept outer wing because it is no longer as parallel to the ground and, therefore, has lower aerodynamic lift. Coincidentally the opposite, or down-swept wing, is more parallel to the ground, increasing its aerodynamic lift. The two concurrent and complementary shifts in wing lift act to rock the bird sideways in the opposite direction, allowing the bird to self-right itself in small-scale, boundary layer turbulence with little, if any, muscular action. Aerodynamically, the back-and-forth rocking flight of Turkey Vultures, as well as that of other dihedral fliers, is far from being uncontrolled and lazy. Rather, it reflects well-formed and adaptive stabilizing, energy-efficient flight. Turkey Vultures in particular benefit from such contorted soaring because they use olfactory cues to detect carrion. Low flight over vegetation benefits *Cathartes* vultures in two ways: first, because small-scale, sheer-induced turbulence is a common and almost dependable source of atmospheric energy at low altitudes within the boundary layer and, second, because flying close to the ground enhances their use of olfactory cues to locate carrion. Although it has yet to be demonstrated in the field, low flight also may reduce resource competition by decreasing the chances of their alerting Black Vultures to the presence of carrion.

The enhanced searching ability of *Cathartes* vultures results in their being behaviorally parasitized by more socially dominant Black Vultures. But it also allows them to more capably seek out and locate small carrion, which they often consume prior to the arrival of Black Vultures following them. Although the extent to which enhanced searching "costs" *Cathartes* vultures has yet to be quantified; three lines of evidence suggest that it may be substantial. First, Black Vultures occur only where Turkey Vultures occur, whereas Turkey Vultures sometimes occur where Black Vultures do not. Second, on islands where Turkey Vultures occur but Black Vultures do not (e.g., Jamaica and the Falkland Islands), Turkey Vultures expand their use of human-dominated landscapes and are more likely to travel in small groups when searching for food. Third, in areas such as tropical northern South America, where more massive subspecies of Turkey Vultures from North America overwinter, the larger migrants typically drive off the less-massive resident Black Vultures not only from carcasses but also from seemingly preferred habitats, including garbage dumps.

Although death by predation is surprisingly rare in ecosystems, death itself is not. Except in the Old World, where closed forest canopies significantly compromise an effective visual search, the carcasses of most large animals wind up in places that make them immediately accessible to vultures. In the New World, olfactorily enhanced *Cathartes* vultures, and other vultures clever enough to follow them, locate large carcasses even on the floors of heavily forested areas. Although smaller carcasses may not be equally available if an animal dies in confined spaces such as animal burrows and natural cavities, Turkey Vultures sometimes find and exhume such carcasses through smell. Although the proportion of living-to-dead animals typically favors predatory to obligate scavenging birds of prey, the required food-catching adaptations of the latter often confine them to much narrower ranges of potential food items size-wise, and scavengers often feed on carcasses that differ considerably in size. In East Africa, for example, vultures can feed on a 90 lb (40 kg) carcass of a Thomson's gazelle and a 9000 lb (4000 kg) male African elephant on a single day. And in Spain, Griffon Vultures sometimes feed on leftover pieces of chicken at a municipal dump in the morning and a 66,000 lb (30,000 kg) fin whale carcass in the afternoon. In the New World, South American Black Vultures feed on road-killed rabbits and bull cows daily. Thus, overall, avian scavengers have a far broader array of potential dietary items, both size-wise and taxonomically, than even the most generalist bird predators have, and the notion that carrion is a scarce resource has been overplayed. As so has its unpredictable nature.

The social behavior of many vertebrates significantly increases the geographic clustering of carcasses, both in natural and human-dominated landscapes. In Africa, both Egyptian Vultures and Lappet-faced Vultures take the eggs, nestlings, and carcasses of adult flamingos and other colonial nesting birds at breeding colonies. In South America, both Black Vultures and Turkey Vultures often feed on marine-mammal placentas and stillborn pups at coastal marine-mammal colonies, and both Turkey Vultures and Andean Condors often visit colony-nesting seabirds on islands along the nutrient-rich Pacific Coast.

Traditional and modern agriculture, too, act to enhance the predictability of domestic carrion. Transhumant herding, the ancient herding practice of moving livestock from one grazing ground to another seasonally—typically to lowlands in winter and highlands in summer, or to productive grasslands following rains—is a significant case in point. This agricultural practice, which originated in parts of Africa and Eurasia tens of thousands of years ago, works in two ways. First, it brings large numbers of ungulates into close quarters, which concentrates general mortality spatially; and second, it moves livestock long distances, often over rugged terrain, which increases accidental mortality temporally. A relationship between transhumant herding and many scavenging raptors, including Bearded Vultures, Cinereous Vultures, and Griffon Vultures in the Caucasus range of Eurasian Georgia, has been underway for millennia. And the same relationship occurs in the Cantabrian Mountains of northwestern Spain, where biologists Pedro Olea and Patricia Mateo-Tomás have documented predictable spatiotemporal adjustments in the locations of seasonal roosts and breeding locations of Griffon Vultures in response to seasonal shifts in numbers of domestic sheep. Although the speed at which vultures shift foraging sites in response to human herding remains surprisingly unstudied, their speed in reacting to newly established vulture restaurants suggests that it is likely intentional behavior.

Other agricultural practices, too, act to increase the foraging efficiencies of vultures. Non-migratory Turkey Vultures on the Falkland Islands quickly coalesce around active compact birthing paddocks for sheep, where they feed on afterbirth and stillborn young. And the same occurs at large-scale dairy and beef farms during calving periods. Large poultry operations and coastal fishing ports serve as gathering sites for New

World Turkey and Black Vultures throughout the Americas, and the same is true of Hooded Vultures in West Africa. Individuals attracted to such circumstances typically consist mainly of juveniles and subadults, whose wide-ranging food-finding and competitive skills are still developing, and for which human-related subsidies may be particularly significant.

In many cases, these and other human–vulture relationships shape the distributions and, sometimes, the abundances of scavengers. To appreciate the many potential relationships involved, it is useful to define them ecologically. The American ecosystems ecologist Eugene Odum first did so in the 1940s. According to Odum, *neutralism* occurs when neither species population is affected by the presence of the other; *competition* occurs when each species population adversely affects the other; and *mutualism* occurs when both species benefit from and require the presence of the other for their survival. *Protocooperation* is similar to mutualism except that the relationship is not obligatory. *Commensalism* occurs when one species benefits but the other is not affected, and *amensalism* occurs when one population is inhibited ecologically but the other is not affected. Finally, both *parasitism* and *predation* occur when one species adversely affects populations, but at the same time is dependent upon those populations.

Human commensalism and, possibly, limited mutualism between humans and vultures are the relationships of interest here. Ubiquitous human commensals include microbes such as *Streptococcus pyogenes*, which inhabit our lower digestive tracts but typically neither help nor harm the guts of the people they depend upon. The relationship, however, can turn parasitic if such bacteria migrate from the digestive tract to other parts of the body, such as the upper digestive tract, where it causes strep throat. Additional examples of widespread human commensalism and mutualism involve domesticated dogs and cats that most likely followed an evolutionary path from wild, facultatively scavenging species into human commensals because of increased food availability provided by humans, along a pathway that led to their eventual domestication.

Today, several species of both New World and Old World vultures benefit by living in close proximity to humans because of increased food availability, while possibly helping humans in their environments. Historical accounts indicate that populations of Black Vultures have increased both numerically and distributionally in North and South America following European settlement, largely as a result of their ability to feed on human waste. Populations of Hooded Vultures appear to have done the same in West Africa: In regions where commensalism occurs, Hooded Vulture populations are orders of magnitude greater than in other parts of their range.

Less-widely recognized examples of human commensalism, and possibly mutualism, include recent increases in Spanish populations of Griffon Vultures and, until recently, populations of three species of *Gyps* vultures in southern Asia. In all such instances, commensalism acts to increase the predictability of carrion and organic waste, spatially, temporally, or both.

A good deal of carrion predictability also occurs in natural landscapes, where its impact is either assumed or is simply overlooked by field workers, and as such is little discussed in the scientific literature. A careful read of the literature, however, documents considerable evidence of carrion predictability and its impact on vulture food-finding behavior in natural landscapes. Several examples come from the savannas of East Africa, where Glasgow University's David Houston studied the scavenging ecology of vultures for several decades beginning in the 1960s. Houston's detailed observations of the breeding behavior and ecology of a colony of Rüppell's Vultures nesting in the Gol Mountains of northern Tanzania, east of Lake Victoria, in the late 1960s and early 1970s are worthy of note.

Parental vultures and condors typically enhance reproductive success by timing their breeding to coincide with predictable peaks in food availability, which is exactly what the vultures Houston studied did. Breeding pairs timed their reproductive efforts so that the most critical period for carrion availability—the time when young left the nest and began food finding on their own—coincided with peak mortality of their principal source of carrion, including migratory wildebeest and zebra, that were dying in elevated numbers at the end of the regional dry season. For Houston's vultures this meant egg laying in December through early February, hatching in late May through early July, and fledging in late July through early October at the end of the dry season, when migratory herds were in marginal condition and mortality was high.

In addition to seasonally timing their breeding activities, the birds timed their daily foraging flights to enhance the likelihood of finding carrion. In East Africa, ungulate carcasses are most available shortly after dawn; both because of nocturnal predation the previous night, and because of overnight accidental and other natural deaths. Not surprisingly, the birds Houston watched departed their colonies as soon as low-cost soaring flight was possible in the early morning. In the Gol Mountains this occurred soon after first light thanks to predictable updrafts created by predominant easterly winds that provided continuous updrafts over cliff faces at their colonies, which allowed the birds to depart their nests by approximately 06:30 am, long before thermals developed later in the day. Vultures that had found ungulate herds the previous day returned to them directly, whereas individuals that had not were more likely to follow the former, with the breeding colony functioning as an information center for individuals in search of new carcasses.

Houston monitored the sizes of the visibly distensible crops of vultures returning to the colony each day and used this information, together with data from the size of visible crops of captive birds fed known amounts of carrion, to assess individual rates of prey delivery to young over the breeding season. This method allowed him to determine if free-ranging adults were able to provide enough food to their developing young. The short answer was yes, but at a measurable cost in adult weight loss and body condition over the course of the breeding season. Even so, 86% of all eggs laid in 1970 hatched, and 96% of the resulting nestlings fledged, indicating that adults were able to successfully reproduce even in the face of measurable food stress. They did so presumably by anticipating predictable seasonal and daily peaks in carrion availability, and by timing their breeding efforts accordingly.

Unfortunately, Houston's detailed study extended across only two breeding seasons, including one in which observations began only after egg laying had begun, and it was not possible for him to determine the extent of interannual variation in seasonal timing of breeding at the colony. Fortunately, a less-intensive, long-term study of the breeding biology of Rüppell's Vultures at a colony site in nearby southern Kenya was carried out in 2002–2009. Although breeding success was not monitored at the colony, the study documented considerable interannual variation in the timing of peak breeding at the site, along with considerable differences in the numbers of active nests, the latter ranging from fewer than 30 in 2004 to more than 60 in 2008, while demonstrating a significant inverse relationship with rainfall during the previous year. Variation in rainfall substantially affected populations of native ungulates in the region, with low rainfall linked to higher mortality and, presumably, higher carrion availability. That individuals were more likely to nest when carrion availability was high, and that the timing of their breeding shifted among years, suggests that Rüppell's Vultures were able to anticipate seasonal shifts in both the magnitude and the timing of carrion availability. The enormous body of long-term ecological knowledge available to researchers working in the Serengeti ecosystem, together with the known geography mortality hot spots within the system, and the fact that five species of vultures currently breed and feed there, make it an ideal location to conduct similar investigations.

Landscape features, both natural and human associated, also sometimes act to increase carcasses' predictably. Beaches and the shorelines of both rivers and lakes that create drift lines of carrion, including beached marine mammals and fish—often in conjunction with seasonal and meteorological pulses—are well-known attractions for searching avian scavengers. In the 19th century, California Condors reportedly gathered by the dozens during the spring at waterfalls along the Columbia River to feed upon dead and dying salmon and trout made vulnerable during freshwater spawning runs. Beached whales, too, were likely used by the species coastally, and although documented accounts are limited, hundreds of condors reportedly fed upon the carcasses of sea lions killed during the pre-petroleum Californian "seal rush" in the 1850s. In South America, stable-isotope analyses of the mid-19th and early 20th-century diets of coastal-nesting Andean Condors, many of which now depend on introduced domestic and wild ungulates in relatively distant Argentinian steppes, once depended upon more proximate, but no-longer abundant marine-mammal carcasses including those of sea lions and whales along the Pacific Coast beaches of southern Patagonia.

Roadways and vehicle traffic, too, increase the spatial predictability of carrion, especially when coupled with increased populations of local wildlife. In eastern North America, mid-20th-century resurgent populations of White-tailed Deer and milder winters, together with expanding networks of roadways and vehicle traffic, have fostered range expansions of both Black and Turkey Vultures in the region. National Audubon Society Christmas Bird Count data suggest that at least some partially migratory Black Vultures conduct mid-distance northern forays back into their breeding ranges coincidental with the arrival of predictable gut piles of hunter-killed deer from large-scale early winter recreational hunting. Roadside counts in northern Patagonia indicate that Black Vultures are more likely to search for and feed upon road kills than on carcasses farther from roads.

Exceptionally severe winter weather that kills large numbers of wild ungulates and livestock also serve to provide predictable carcasses.

"Dzuds," or particularly harsh Mongolian winters, such as the one in 2009–2010 that killed millions of ungulates, shift the migratory behavior, feeding ecology, and, presumably breeding success of Cinereous Vultures, many of which feasted on hyper-abundant frozen and thawing livestock carcasses.

Large-scale spikes in ungulate mortality during seasonal migratory movements, too, can increase the predictable nature of the carrion landscape. Numerous migrating zebras and wildebeest predictably drown while attempting to cross the crocodile-infested Talek River in the Maasai National Reserve in southwestern Kenya, where the seasonal appearance of floating and semi-submerged carcasses plays host to large numbers of vultures that feed on them.

SEARCHING FOR FOOD

With body masses that in some species exceed 26 lb (12 kg), vultures rank among the largest of all flying birds. Indeed, even the mass of the smallest vultures almost always exceeds 2 lb (1 kg). Mathematical models suggest that large body size is selected for in vultures because carrion, particularly that of large animals, is often available only episodically. Individuals need to be able to survive on large-body fat reserves between periods of food finding. That said, carrion size is frequently related to body size, with smaller species, including Egyptian Vultures, Hooded Vultures, Black Vultures, and Turkey Vultures feeding on smaller carrion, and larger species, including Cinereous Vultures, Griffon Vultures, and Cape Vultures, feeding on larger carrion, overall. Kleiber's law, named for the Swiss biologist Max Kleiber for work in the early 1930s, indicates that the basal metabolic rate of a vulture scales to a ¾ power of the individual's body mass, as well as to the extent of its use of low-cost non-flapping soaring versus higher-cost powered flapping flight. Consequently, large-bodied and largely soaring scavenging raptors have relatively low basal metabolic rates, even when compared with other birds of similar size.

In addition to large size and an adaptive capacity for existing on episodic, large-meal bouts of feeding, vultures have two additional adaptations that enhance the likelihood of an obligate scavenging life: low-cost soaring flight and both within- and among-species social facilitation when food searching. The keen sense of smell in several New World species also helps. Low-cost soaring flight permits metabolically efficient food searching behavior, and social facilitation enables several New World species to find carrion in heavy vegetation, including closed-canopy forests.

Soaring Flight

Many vertebrates, including fish, amphibians, reptiles, and mammals, routinely glide but only birds routinely soar. Among the birds, only a few soar as often as vultures, or depend on soaring as much. Non-flapping gliding and soaring flight, which are powered by external, non-physiological energy sources, provide low-cost aerial transport. Of importance, both allow individuals to move across landscapes while searching for carcasses and co-conspirators in ways that have enabled this group to accomplish what other vertebrates have not: to thrive exclusively on carrion.

Avian soaring behavior has attracted human attention for centuries, not only out of curiosity but also because of its potential in human aviation. Having developed an understanding of atmospheric movements in the 1920s, meteorologists quickly laid out principles and mechanics of avian soaring flight in birds. By the early 1970s, British scientist and power-glider pilot Colin Pennycuick was applying aerodynamic principles to explain the effective nature of cross-country gliding and soaring in New and Old World vultures. His benchmark work remains foundational even today, and I summarize several aspects of it below.

Pennycuick was one of the first to recognize that soaring flight was a necessity in the group because vultures and condors were incapable of sustained flapping flight due to their large size, which is a function of allometric scaling. Power required for flapping flight in birds increases more quickly than body mass, meaning that this type of flight sets an upper limit on a bird's ability to fly using internal metabolic power. Because of this relationship, many large birds, including vultures, storks, and pelicans, near the upper limit of this relationship are not able to remain aloft via prolonged flapping flight and, as a result, need to use atmospheric energy to do so. Gliding flight, which depends on the use of potential energy associated with gravity and an individual's height above the ground, allows scavenging raptors to take off from elevated perches and glide down while generating lift on outstretched wings. But unless the air is moving upward at a rate greater than the downward movement of the gliding bird, vultures need to use flapping flight to move forward so as to generate the aerodynamic lift needed to keep them aloft. And that is exactly what occurs when vultures depart from tree-top or cliff-side perches each morning. They deftly maneuver in flapping flight while seeking pockets of the upwardly moving air they need to supply atmospheric updrafts for soaring flight. Pockets of upward moving air are critical for successful soaring flight, but vultures must also know where to find it, and once found, they must be able to operate successfully both within and among such airspace as they travel across the landscape. Two types of atmospheric updrafts are available over land: slope and vegetation updrafts, and thermals, and both are used by vultures. Slope and vegetation updrafts are more common in topographically undulating areas where hills, mountains, and cliffs create updrafts as horizontal winds are lifted up and over such barriers, and in mixed open and arboreal habitats where isolated woodlands do the same. Slope updrafts are particularly useful shortly after daybreak at a time when differential heating of the earth's surface, and the thermals it produces, have yet to form. Later in the day, thermals become increasingly important.

Thermals are pockets of warm, rising air that form when different surfaces in landscapes receive and absorb different amounts of sunlight. Dark surfaces have lower reflectivity indexes, or surface albedos, than do light-colored surfaces, and as such absorb more solar radiation than the latter. Dry surfaces, including rocky outcrops and parched fields, heat more quickly than wet surfaces, such as water and living vegetation. In hilly areas, surfaces that are oriented perpendicular to incoming solar radiation warm more quickly than do shaded and less-perpendicular surfaces.

Land-based thermals typically require sunlight to sustain them. Thus, the strongest and largest thermals generally occur on bright, sunny days between midmorning and midday, after the sun is high enough to differentially warm the landscape and create thermals, but before strong afternoon horizontal winds begin to tear them apart. Because land-based thermals are fueled by sunlight, thermals are stronger in summer than in winter in non-tropical areas, and are stronger overall in the tropics.

Thermals are most useful to vultures when they are wide enough for raptors to circle within them, and when they are strong enough to allow individuals to reach heights sufficient to glide in the intended direction of migration to the next thermal. Oceanic or sea thermals also occur. And although they play a critical role in shaping the geography of long-distance raptor migration in several predatory species, they do not apply here.

Students of soaring flight often portray thermal soaring as a series of low-speed, upward, corkscrew-like, circular soaring flight as birds encounter small- to medium-sized discrete pockets of warm, rising air interspersed with slowly descending gliding flight between thermals. However, such flight represents one end of the spectrum of soaring techniques and is not necessarily the most common. In many areas, vultures glide and soar through landscapes in which so-called thermal streets and mountain-ridge slope soaring play major roles. The soaring behavior of *Gyps* and other vultures in the Serengeti of East Africa offers a good example of this far-more complicated soaring flight practiced by avian scavengers there and elsewhere.

Until recently, as many as six species of vultures coexisted in this wooded savanna ecosystem, with assemblages of all six sometimes feeding simultaneously on the same carcass. The two most common species in this scavenger community, Rüppell's Vulture and White-backed Vulture, feed mainly on the muscles, guts, and other soft innards of large carcasses. Two additional species, the White-headed Vulture and the Lappet-faced Vulture, eat mainly skin, sinews, and muscle tissue adhering to bones at such carcasses. The two remaining species are the far-smaller Hooded Vulture and the Egyptian Vulture, although the latter is currently all but absent in the region. Both of these have slender bills and feed mainly on small fragments of meat left by the other larger vultures. The breeding and related feeding ecology of the largely non-territorial Rüppell's Vulture, and to a lesser extent the White-backed Vulture, are strongly linked to the movements of migratory wildebeest and zebras in the region. By comparison, the White-headed Vulture and Lappet-faced Vulture, both of which are territorial, are more eclectic and somewhat less dependent on the movements of wildebeest and zebras. Hooded Vultures depend on the larger vultures to provide the tidbits they typically consume at carcasses; on humans, who often provide them with food; and on live insects associated with large carcasses. All six species travel long distances when searching for carcasses, and all piggyback on the searching behavior of other species to help them locate carcasses. Although soaring behavior sometimes differs among species, all employ the same general aerodynamic rules when searching for carcasses.

While following Rüppell's Vultures in a powered glider, Pennycuick found that during good soaring conditions (i.e., relatively sunny and rainless days during the region's dry season), most birds maintained speeds of approximately 28 mi^{hr} (45 km^{hr}). Individuals traveled 2 to 3 hours between their cliff-face nesting sites and the feeding areas used by migratory ungulates. They took off from nest sites at first light and slope soared along cliff faces and hill sides until they encountered emerging thermals later in the day. Under such conditions, climbing in thermals averaged between 6 and 12 ft^{sec} (2–4 m^{sec}) with straight-line glides of 45–55 mi^{hr} (70–85 km^{hr}) which translates into an overall cross-country speed of 26 mi^{hr} (43 km^{hr}). Individuals typically flattened their glide ratios and slowed their forward speeds when gliding through thermals, and sped up their flight by flexing their wing to increase their wing loading, and hence their speed, between thermals.

Overall, species with lower wing loadings, including Lappet-faced Vultures and White-headed Vultures, were able to fly earlier in the morning on smaller and weaker thermals, whereas as those with higher wing loading, including Rüppell's Vultures and White-backed Vultures, were able to glide more rapidly between thermals. As a result, the first two began food searching earlier each day and stayed aloft longer, whereas the latter two engaged in more rapid cross-country flight overall. Such differences suggest that the former might be at a competitive advantage over the latter. However, observations in the Serengeti indicated that Rüppell's Vultures sometimes roosted nocturnally near the herds during the dry season, but not other seasons, and that they, together with White-backed Vultures, which nest closer by, actually outnumbered the Lappet-faced Vultures at early morning carcasses. In addition, Rüppell's Vultures had higher feeding successes—as evidenced by proportions of individuals seen with full distensible crops then—than did the Lappet-faced Vultures, presumably because of their more aggressive gang behavior when feeding, and their greater use of socially facilitated searching networks. Based on extensive in-flight observations, Pennycuick concluded that most individuals were quite good at finding the atmospheric updrafts, including thermal streets oriented in or near the direction of their intended travel, by watching for and moving to align themselves under lines of cumulus clouds that typically formed above such atmospheric phenomena. Flying in thermal streets increased cross-country travel speed considerably.

Although the principal energy source of soaring flight is external, there also is an internal metabolic expense. The pectoral, or breast, muscle of vultures holds the outstretched wings in place at such times, and in all soaring vultures the pectoralis is subdivided to include a band of tonic, or slow muscle, that acts to sustain the necessary tension. The actual metabolic cost, although likely negligible, remains largely conjectural. Heart-rate monitors attached to vultures using this rate as a proxy for metabolic expense during soaring versus other activities, such as "alert perching" and flapping flight, indicate that the heart rates of free-flying Himalayan Vultures and Griffon Vultures were lowest in early morning when the birds were at rest, approximately twice as high later in the day when birds were perched but vigilant, and about three times higher still when the birds engaged in flapping flight. Significantly, heart rate returned to close to the alert perching-but-vigilant baseline when soaring, with individuals returning to that baseline within several minutes after ceasing flapping flight. The heart rate of a similarly monitored Turkey Vulture indicated considerable variation in heart rate when perching, with no significant overall increases in hourly heart rates during soaring flight, regardless of hourly ground speeds of travel ranging from less than 6 mi^{hr} (10 km^{hr}) to as much as 18 mi^{hr} (30 km^{hr}).

Wing Shape and Size

The shape and size of a bird's wing, and its influence on aspect ratio (i.e., the ratio of wingspan to the mean chord [width] of the wing) and wing loading (i.e., the bird's body mass divided by the combined areas of its outstretched wings), have long attracted the attention of students of soaring flight, as has the occurrence and distribution of atmospheric updrafts. Most over-land soaring specialists, including eagles, storks, cranes, and vultures, have largely rectangular wings with low aspect ratios of 8 to 10, with slotted wingtips with finger-shaped or emarginated primary feathers, whereas over-water soaring specialists, including albatrosses and frigate birds, have narrower wings with higher aspect ratios of 14 to 16. According to Pennycuick, the difference in shape lies not so much in gliding or soaring performance, or to differences in the use of thermal versus dynamic (aka wind-shear) soaring in the two groups, but rather to differences in flight requirements during takeoffs and maneuvering in vegetation. Feeding vultures—excepting those feeding on steep slopes—need to accelerate to a minimum power speed required to take off in level flapping flight, something that aerial feeding marine birds do not need to do. Feeding on level ground and taking off in vegetation typically requires considerable flapping flight, and ground takeoffs have substantial physiological costs. Observations of *Gyps* species in East Africa, for example, indicate that taking off from a carcass with a full crop often requires sustained bouts of 40 to 60 or more flaps before reaching heights needed for soaring flight. Add to that the so-called dread flight undertaken by individuals and groups of perched vultures frightened by approaching ground predators, and it becomes apparent that on-the-ground feeding has considerable metabolic implications.

Initially developed as an aeronautical term in the early 1900s, wing loading, or the wing area of an airplane divided by its mass, was adopted as an anatomical measure by avian biologists in the 1930s. Frequently characterized as species-specific, unlike its use in fixed-wing aircraft, avian wing loading is dynamic and changeable within species because birds frequently flex or crimp their wings in gliding flight—increasing their wing loadings accordingly—depending on the immediate task at hand. Usually expressed in lb/in^2 (kg/m^2), wing minimal loading on fully outstretched wings is a widely used metric in avian gliding flight in part because it influences the range of air speeds that a bird can travel at when gliding, as well as its minimal turning radius when soaring in thermals. Light wing loading allows a bird to glide on outstretched wing at lower speeds without stalling and allows it to soar in tighter circles in thermals. The latter being true in part because circle soaring often requires a steeper banking angle that reduces the effective size of the wings in generating lift. As a result, lighter wing loading increases the efficiency of both searching and cross-county soaring flight and, not surprisingly, vultures have some of the lightest wing loadings of all large soaring birds. The literature is full of discrepancies between observed versus calculated values regarding the extent to which wing loading affects gliding speeds and turning radii of individual species. Why? Because 1) scavenging raptors and other large soaring birds do not always fly on fully outstretched wings, and wing area changes substantially with wing postures; 2) body mass varies considerably both within and among individuals; and 3) field workers sometimes measure wing areas in dissimilar ways. These discrepancies challenge our ability to assess the extent to which wing loading affects soaring ecology. In spite of this, several seemingly adaptive associations between wing loading and flight behavior are apparent.

Relatively large-bodied, mountain and cliff-nesting species, including Bearded Vultures, Griffon Vultures, Rüppell's Vultures, Cape Vultures, Himalayan Vultures, California Condors, and Andean Condors, all of which have relatively high wing loadings, frequently slope soar at high speeds to generate additional lift needed to remain aloft across long distances.

Wing loadings also can differ within species based on population-specific differences in movement ecology. Argentine vulture specialist Maricel Graña Grilli recently compared wing loadings of satellite-tracked migratory Turkey Vultures from western Canada, southern Arizona, and northern Patagonia with those of partial migrants from eastern Pennsylvania and wing-tagged non-migrants from the Falkland Islands. Graña Grilli's team found significant differences in wing loadings between migrants and non-migrants, and between long- and short-distance migrants, with migrants from populations with lower wing loadings flying farther, faster, and higher during their migrations than those with heavier wing loadings, indicating that in this species at least, long-distance migration is linked to lighter wing loading. It most likely allows individuals to soar higher in thermals and to glide more efficiently between them, thereby enabling Turkey Vultures to speed their migratory travels.

Social Networks

Although soaring provides an energy-efficient searching mechanism for vultures and condors, that alone is not necessarily sufficient for successful food finding. What the African vulture specialist Peter Mundy once characterized as "information networks in the sky" also play a critical role in the hyper-efficient food finding abilities of scavenging raptors. While scientists studying predatory lions and hyenas were among the earliest to popularize the critical importance of aerial information networks in the 1960s, the English ornithologist Henry Tristram is credited with the first description of the phenomenon. Writing in 1859, Tristram described the avian search network as follows:

> The griffon who first [discovers] his quarry descends from his elevation at once. Another, sweeping the horizon at still greater distance, observes his neighbor's movements and follows his course. A third, still further removed, follows the flight of the second and is traced by another; and so, a perpetual succession is kept up as long as a morsel of flesh remains over which to consort. I can conceive no other mode of accounting for the numbers of vultures which over the course of a few hours will gather over a carcass.

Reports of such networks acting to assemble huge, albeit sometimes apocryphal, numbers of scavenging raptors date to the 8th century when an epic poem, *The Song of Roland*, detailed a mountain-pass confrontation in the Pyrenees between Charlemagne and local Basques that attracted an assemblage of some "30,000 thirsty grifins [*sic*]" that feasted on the equine and human remains. More recently, Arabian tales of thousands of vultures coalescing at Crimean War battlefields, and newspaper accounts of masses of Turkey Vultures feasting on equine and human remains at the Civil War battlefield of Gettysburg, Pennsylvania, confirm the phenomenon. And indeed, anthropologists have suggested that sub-Saharan vulture information networks were being used by early hominin hunter-gathers more than 100,000 years ago to follow vultures to carcasses, just as hyenas and jackals do today.

Although socially facilitated food finding has been known for some time, its efficiency has proved somewhat difficult to measure in the field. Nevertheless, it has been modeled. In 2008 a team of scientists from Ireland and Scotland used an individual-based, spatially explicit simulation model to link foraging success with densities of carcasses and searching birds. Using field data from the Serengeti ecosystem, the team proposed that increased numbers of aerial foraging individuals enhanced the impact of social facilitation on food finding across a wide range of carcass–vulture densities seen in the field, suggesting as part of their argument that single soaring vultures were able to detect a carcass accompanied by a feeding group of vultures from distances of up to 2.4 mi (4 km), and unoccupied carcasses from distances of up to 325 yd (300 m). The relationship was best described by a sigmoid, or s-shaped, curve that depended on the number of carcasses, with increased numbers of carcasses exponentially increasing the likelihood of food finding until the total number of searching vultures saturated the system. The researchers concluded that social facilitation was extremely efficient, and that it was most likely responsible for hundreds of birds reaching a carcass within hours of its appearance. That efficiency results in vultures being the dominant secondary consumers in African savanna ecosystems, well ahead of large-cat predators. Their results indicate that efficient food finding occurs only at vulture densities great enough so that all individuals are within sight of other individuals, and that once populations fall below a critical level, foraging success drops dramatically. The latter point being of considerable importance for vulture conservation. Although similar modeling has yet to be applied to vulture searching success elsewhere, numerous observations in Europe, Asia, and the Americas suggest that social facilitation in searching occurs there as well.

Within the Americas, social facilitation is complicated by the fact that three New World species, the Lesser Yellow-headed Vulture, Greater Yellow-headed Vulture, and Turkey Vulture, have keen senses of smell that allow them to detect carcasses by smell as well as by sight, and that four other species in the New World at least occasionally follow the former species to carcasses. Black Vultures, in particular, are known to monitor the former species by tracking their activities while soaring at greater heights. Although such "parasitic" behavior has been recognized for some time in North America, fieldwork in Central and South America indicates that it occurs there too.

Almost certainly, Andean Condors and California Condors, both of which typically appear at carcasses after Turkey Vultures, follow the Turkey Vultures, as do King Vultures. King Vultures have more massive and stronger bills than *Cathartes* vultures and are better equipped to open larger carcasses, and some researchers have suggested that the smaller-billed species sometimes seek out the King Vultures and lead them to large carcasses so they can open them, making them available to all. Although this mutualistic relationship is yet to be documented in the field, observations of Turkey Vultures at several zoological gardens support the possibility. Several years ago, when I visited two gardens in Venezuela as possible sites for wing-tagging hundreds of overwintering Turkey Vultures that roosted there, I was struck by the fact that most of the free-ranging migrants roosted disproportionately in trees directly above aviaries containing King Vultures, but not those containing Andean Condors—seemingly waiting upon the former to follow them to large carcasses.

Additional evidence for the extent to which social facilitation affects foraging success in vultures comes from observations of New World Turkey Vultures at the northern limits of their distributional range in western North America. American field biologist Marc Bechard and I assessed the distribution and abundance of populations breeding largely in abandoned buildings along more than 1865 mi (3000 km) of road surveys in south-central British Columbia, Saskatchewan, and Manitoba, Canada, and in north-central Minnesota, USA. Turkey Vulture densities averaged fewer than 0.3 individuals per 60 mi (100 km) along our survey routes, or less than 10% of the densities typically encountered well within the species range farther south. Although we expected the low densities, their distribution along the routes surprised me. Rather than being somewhat evenly spread throughout, isolated pockets of small groups of Turkey Vultures occurred episodically along short, <6 mi (10 km), stretches of the routes, usually in riparian areas or near small towns that were separated by stretches of more than 60 mi (100 km) where no birds were seen. The geographic groupings we observed certainly increased the likelihood of isolated aerial food-finding networks in this region of low population densities. Support for this comes from a study of the breeding ranges of six individuals summering in a breeding area in south-central Saskatchewan, again near the northern limits of the species range. This investigation indicated that the home ranges of these individuals averaged no greater than that of similarly tracked individuals from southeastern Minnesota, Indiana, Ohio, and South Carolina, USA, all of which were breeding in denser populations well within the species distribution range. Although the extent to which the Canadian breeders used aerial networks was not determined, their use of similarly sized home ranges of birds breeding to the south suggests they were searching for carcasses similarly, and that clustering breeding in presumed ecological hotspots allowed them to use isolated aerial networks to enhance their carcass-searching success.

In addition to following others to locate food, vultures sometimes follow mammalian predators that are hunting in anticipation of prey capture and carcass availability. During six seasons of fieldwork in Venezuela, French vulture specialist Marsha Schlee watched King Vultures routinely track hunting jaguars during 27 of 162 days of observation, with vultures feeding at kills after the jaguar had finished feeding and left the area. Historically, legendary German naturalist Alexander von Humboldt reported Black Vultures behaving similarly in Venezuela in the late 1700s. More recent studies indicate that Andean Condors follow pumas that are hunting guanacos in southern Chile. In the Old World, hyena specialist Hans Kruuk watched Hooded Vultures doing the same while tracking the movements of food-searching hyenas. Kruuk reported Hooded Vultures showing up even before hyenas when he played tape recordings of bone-crunching hyenas on a kill, suggesting that this vulture uses sound as well as sight to locate food.

Wildfire also informs vultures of food availability. Accounts from the 18th century document the arrival of California Condors or Turkey Vultures, or both, at riverside fires set by Native Americans during festivals associated with salmon runs in the American Pacific Northwest along the Klamath River in northern California, USA; and in Africa, brush fires attract Palm-nut Vultures who feed on the carcasses of burned reptiles and amphibians.

Aggressive Scavenging

In addition to being attracted to areas where animals are likely to have died, field observations suggest that vultures sometimes attack and feed particularly distressed and vulnerable live prey before they die. Although instances of aggressive or opportunistic scavenging have been described by both farmers and field biologists, actual observations of vultures or condors attacking, killing, and eating sick or seriously injured wildlife and livestock are relatively uncommon. Such behavior likely occurs only when other nutritional resources are unavailable. African vulture specialist Peter Mundy reports two cases involving captive Cape Vultures, one in which an individual attacked another having a seeming epileptic seizure, and a second in which a captive individual attacked a Black Eagle entangled in aviary fencing. During six seasons of field observations in the Falkland Islands, an archipelago in which domestic sheep outnumber humans, I witnessed two events in which invalided, old sheep that had fallen down and were unable to stand up were approached by Turkey Vultures and fed upon while still alive. I also saw Turkey Vultures feeding on a ewe that had fallen into a narrow, steep-sided creek and was standing submerged to its shoulders.

Cannibalism

Cannibalism routinely occurs in obligate siblicidal predatory raptors where nestlings kill and consume siblings when food is scarce. However, it appears to be relatively rare among vultures. Observations include an ambulatory nestling Griffon Vulture feeding on a dead adult at a nest; two young Griffon Vultures feeding on a dead conspecific killed by a poisoned carcass; several Griffon Vultures pecking at, and subsequently killing and eating, a conspecific trapped under a rock while feeding on a dead sheep; and a parental Bearded Vulture, a known siblicidal species, feeding the remains of a young chick to its older sibling at a nest. In Africa, Peter Mundy and colleagues reported *Gyps* vultures feeding on the corpse of an African White-backed Vulture snagged by its wing in an elephant's carcass. New World examples include several events involving captive King Vultures, and a record of conspecifics feeding on dead Lesser Yellow-headed Vultures at a road kill in Mexico. American ornithologists Noel Snyder and Helen Snyder reported that placing Turkey Vulture carcasses at vulture feeding stations in Florida strongly inhibited the use of such sites by others. And in coastal Veracruz, Mexico, where the flight of approximately 3 million Turkey Vultures occurs each autumn, dead migrants are not consumed by others. I observed two Turkey Vultures killed during a winter storm in the Falkland Islands that were fed upon by several Striated Caracaras and at least one Variable Hawk, but not by several Turkey Vultures that flew low above the clearly visible carcasses.

COMPETITION AT CARCASSES

Competition among Vultures

The aerial information networks that help vultures locate carcasses are a two-edged sword. Although they help individual birds find carcasses far more quickly than otherwise, they also bring large numbers of hungry vultures together, and in so doing they create situations in which there is not enough food to go around for all involved. The conflagrations that result create significant, and often daily, struggles for those involved.

Scientists consider interspecies competition to be among the most ubiquitous of all ecological interactions. Indeed, as the noted British ecologist and competition specialist David Lack once said, "seeing up to six species of vultures feeding on the same carcass on the Serengeti plains . . . one might think that there was an overabundance of food, and no effective competition. This is wrong, however." As is true of biodiversity in general, species diversity in scavenging raptors owes its origin to competition. Whereas intraspecies competition often is more important in affecting the abundance and distribution of scavenger populations, interspecies competition, too, can play a significant role.

Ecologists recognize two types of interspecies competition. *Exploitative* competition occurs when an individual negatively affects another indirectly by acquiring a limited resource in a way that deprives others from acquiring it. *Interference* competition occurs when an individual negatively affects another individual directly by preventing access to a resource through aggressive behavior. The bone-crunching behavior of the more massively billed Lappet-faced Vulture, which allows the species to rip off attached tendons and sinews and crack open skeletal elements to access encased marrow and brain matter, typifies exploitative competition. Scrums of *Gyps* vultures interfering with the feeding efforts of others typifies interference competition. Many who have studied the actions of the six-species guild of East African vultures feeding simultaneously at large carcasses have noted that the six species involved tend to be divided into three pairs of species that differ in the manner in which they feed and the nutritional resources they depend on. Here I describe the significant differences and similarities involved while attempting to explain how they, together with the episodic superabundance of carrion at large carcasses, maintain both this diversity in Africa, as well as the diversity of scavenging communities in southern Asia and the Americas.

One of the earliest accounts of interspecies competition in African vultures was a short note written by the American wildlife ecologist George Petrides published in *The Auk* in 1959 that described interactions among five species in Uganda. Petrides was the first to categorize the order in which different species landed at carcasses, with Hooded Vultures being first, followed by White-backed Vultures and Rüppell's Vultures, and, then Lappet-faced Vultures and White-headed Vultures. He also noted that Hooded Vultures largely retired to the fringe of feeding assemblages as arriving *Gyps* greedily sparred among themselves while tearing apart and swallowing pieces of the carcass; but Hooded Vultures stayed longer, picking up tiny bits of the remains. Petrides also reported that the most numerous and—presumably most successful—species in the area, the White-backed Vulture, was not the largest and fiercest species. He was also the first to ask whether species that arrived early did so because they were abundant, or because they were better at finding carcasses more rapidly. Petrides concluded his descriptions by noting that a detailed appraisal of interspecies competition among guild members would require additional observations of the feeding and breeding habits of the same species in additional locations with different levels of competition.

Dutch ecologist Hans Kruuk, working together his wife, Jane, was the next to describe competition among East African vultures, this time among six species feeding on migratory ungulates in the Serengeti Plains of Tanzania. Kruuk, writing in the mid-1960s, concluded that the two species of *Gyps*, the White-backed Vulture and Rüppell's Vulture, fed mainly on "the carcass itself, pulling out and swallowing the soft and fleshy parts," and that Lappet-faced Vultures, White-headed Vultures, Hooded Vultures, and Egyptian Vultures fed on other parts of the carcass, in addition to the soft innards. He also noted that the long necks of *Gyps* allowed them to reach far inside large corpses, and that their "large, broad and gutter-shaped [tongues were] lined along the edges with long horny 'teeth,' pointing inwards and backwards," thereby increasing their ability to pull out and swallow masses of soft tissue. Kruuk mentioned that Hooded Vultures, White-headed Vultures, and Lappet-faced Vultures tended to be early arrivals, followed by the near simultaneous arrival of the two *Gyps* species; that most aggressive encounters occurred intra-specifically; that Lappet-faced Vultures were the most likely to fight, followed by the two *Gyps*; and that White-headed Vultures were the least likely to fight. Furthermore, that interspecies encounters between Rüppell's and White-headed Vultures were usually initiated by the former. Kruuk also pointed out that Hooded Vultures were less likely than other vultures to be bothered by the presence of mammalian predators; that smaller Hooded Vultures and Egyptian Vulture remained on the edges of scrums at crowded carcasses, where they focused mainly on taking small scraps of carcasses and on insects; and that both species were more likely than the others to be found feeding around human settlements.

David Lack suggested that differences such as these demonstrated that the "various species [were] at least partly separated from each other [via] different specializations," which functioned adaptively to reduce interference competition. Finally, like Petrides before him, Kruuk concluded that additional insights into the impact of interspecies competition in the guild would come only on the heels of additional studies.

More recent studies have focused on observations of species-specific feeding behavior at experimental carcasses and vulture restaurants, as well as on the searching behavior of satellite-tracked individuals followed to assess the extent of interspecies competition. In one such investigation in the Maasai Mara National Reserve in southern Kenya, the American vulture specialist Corinne Kendall noted the behavior of Lappet-faced Vultures, White-backed Vultures, and Rüppell's Vultures feeding together at 2 kg experimental carcasses of the head, organs, and leg of a sheep or goat to assess species differences in timing of activity patterns and foraging success as quantified by the fullness of the birds' distensible crops.

Contrary to predictions that the earlier rising Lappet-faced Vulture would arrive at carcasses earlier in the day before strong midday thermals had formed, the two *Gyps* species were more abundant at carcasses in morning and Lappet-faced Vultures were more common in afternoon, even though the latter were more common on morning transects. Kendall also noted that crop fullness was greater in Rüppell's Vultures, with 79% of roosting individuals having fully distended crops, less so in White-backed Vultures at 59%, and lowest in Lappet-faced Vultures at 29%. Fullness peaked in White-backed Vultures at 09:00 in the morning versus 11:00 in Rüppell's Vultures, and later in the afternoon in Lappet-faced Vultures. Intriguingly, Kendall noted that age did not have an effect on crop fullness in any species. Her results demonstrated that Rüppell's Vultures—a species that earlier studies had shown were most aggressive toward other species at carcasses—filled their crops earlier in the day.

Another study by the Israeli scientist Orr Spiegel and colleagues in Etosha National Park, Namibia, using tracking devices and behavioral observations at natural and experimental carcasses, focused on White-backed Vultures and Lappet-faced Vultures in a region where Rüppell's Vultures did not occur. Lappet-faced Vultures, which were only a fifth as numerous as White-backed Vultures were the first to arrive over carcasses but often seemed reluctant to land, and were twice as likely to land together with White-backed Vultures rather than precede them. Simulation modeling indicated that the spatial arrangement of roosting sites significantly influenced aerial searching efficiency, with the more widely spaced roosting behavior of Lappet-faced Vultures being favored overall. The authors concluded that higher visual acuity in Lappet-faced Vultures, an ability they assumed based on the species 1.4-fold larger eyes, allowed them to locate carcasses more quickly, and that this, along with the species tendency to depart from nocturnal roosts earlier in the morning, led them to find carcasses earlier in the day. They also suggested that the species failure to land at carcasses they arrived at first might be because they were waiting for the White-backed Vultures to land and strip and swallow the soft body parts, making it easier for them to feed on the exposed sinews, tendons, and cranial matter that form much of their diet. That Lappet-faced Vultures varied considerably in terms of their searching behavior in Kenya versus Namibia suggests that the species may have behaved differently because of regional differences in interspecies competition.

Although interspecies competition has been studied less extensively in the New World, several observations merit note. American conservation biologists Michael Wallace and Stanley Temple focused on the feeding and aggressive behavior of Andean Condors, along with those of King Vultures, Black Vultures, and Turkey Vultures, at 217 carcasses in northern Peru in 1980 and 1984, including 53 carcasses for which they recorded the order of individual arrival. Turkey Vultures fed at the carcasses on 83% of the days studied, Black Vultures on 62%, Andean Condors on 52%, and King Vultures on 16%. Turkey Vultures usually arrived at carcasses first, Black Vultures second, and Andean Condors third. King Vultures typically arrived in pairs or what were believed to be family groups, and Andean Condors, which typically were the last to land, sometimes waited for hours or days to do so. Interspecies dominance hierarchies based on 8066 aggressive encounters revealed several consistencies, including that during one-on-one encounters, Turkey Vultures won slightly more than half of their encounters with single Black Vultures but were easily intimidated and subordinate during encounters with groups of the latter, and that more massive King Vultures overwhelmingly dominated both. Similarly, the more massive Andean Condors won all 185 aggressive encounters with King Vultures. When smaller vultures fed together, simultaneous access to the carcass was limited to five or six birds, with subordinate individuals waiting nearby for those with access to be sated. The authors concluded there was a strong correlation between dominance and body mass, but that groups of smaller Black Vultures often dominated larger single Turkey Vultures.

A second, less intensive study of competition in New World Vultures focused on species differences in arrival times of King Vultures, Black Vultures, Turkey Vultures, and Lesser Yellow-headed Vultures at 33 fish and one dolphin carcass that researchers placed in open and forested areas near Pacific Coast Costa Rica during the summer of 1987. The researchers also noted that when carcasses were placed in open areas, both Turkey Vultures and Black Vultures landed on the day of carcass placement, and that King and Lesser Yellow-head Vultures did so a day later. When carcasses were placed in forests, only Turkey Vultures arrived on the first day, and King Vultures and Black Vultures arrived a day later. A 21-day study by American vulture specialists Jack Eitniear and Steven McGehee documented both arrival sequences and aggressive interactions involving Lesser Yellow-headed Vultures, Turkey Vultures, and Black Vultures at researcher-placed fish carcasses near a coastal fishing village in Belize. In that study, Lesser Yellow-headed Vultures arrived first 15 times, and Turkey Vultures arrived first six times. Both were almost always supplanted by an individual or groups of larger Black Vultures, and Yellow-headed Vultures almost always were supplanted by larger Turkey Vultures. Lesser Yellow-headed Vultures have significantly wider bill gapes than do larger Turkey Vultures, a feature that Eitniear has suggested is an adaptation to speed feeding in often earlier-arriving and subordinate species.

Finally, a study in northwestern Argentine Patagonia that focused on Black Vultures, Turkey Vultures, and Andean Condors at naturally occurring carcasses indicated that Andean Condors and Black Vultures fed primarily on the carcasses of domestic livestock whereas Turkey Vultures fed on a wider variety of carrion, including fish, reptiles, and birds as well as wild carnivores and rodents. Groups of Black Vultures usually overwhelmed both Turkey Vultures and Andean Condors at carcasses to the point that the latter often did not land at carcasses when groups of Black Vultures were present.

Perhaps because of its episodic nature, the extent to which food fighting at carcasses results in permanent injuries remains largely unreported in the literature; however, circumstantial evidence suggests that it does. David Houston reported finding healed wing fractures in the long-bone ulnas in three of 17 vulture wings collected during studies of feather molt in both White-backed Vultures and Rüppell's Vultures in northern Tanzania. Although Houston was unable to determine conclusively that the fractures he x-rayed resulted from disputes at carcasses, that 20% of the wings he examined had been broken and subsequently healed is considerable, given that the rates of healed ulnas in other groups of birds, such as ducks, gulls, and feral pigeons, do not exceed 0.5%.

Overall, competition studies suggest that 1) differences in bill structure act to separate several species into different feeding niches based on their ability to open, and subsequently consume part of large ungulate carcasses; 2) differences in distributional ranges reduce competition both within and among species in some areas; 3) within species, juveniles and subadults reduce competition with superiorly competitive older individuals by feeding in nursery areas where adults do not breed; 4) differences in food preferences, including opportunistic predatory behavior, reduce competition among species; 5) both intra-and interspecies competition is greater at large versus small carcasses; 6) intra- versus interspecies competition results in a greater number of aggressive interactions at carcasses; 7) gang behavior enables smaller species to outcompete larger-sized species at carcasses; 8) injuries suffered during food fights can include broken wing bones; 9) differences in soaring abilities and in the New World olfactory capabilities allows some species to locate carcasses earlier in the day, more quickly, and most likely, more efficiently than other species; and 10) territoriality may function to reduce both intra- and interspecies competition in some larger species. In sum, competition for food at carcasses is arguably as dynamic and complex in vultures as it is in other groups of organisms.

Competition with Non-vultures

Other birds. As mentioned in Chapter 1, many predatory raptors, including many eagles, are facultative scavengers, several of which are noteworthy regarding their interactions with vultures at carcasses. In East Africa, both Steppe Eagles and Tawny Eagles typically arrive first at early morning carcasses, either because these eagles' relatively smaller size allows them to use weak early morning thermals compared with larger and more heavily wing-loaded vultures, or because their superior visual acuity allows them to better locate carcasses. Although both species are routinely chased from carcasses by later arriving groups of *Gyps*, their predatory beaks allow them to slice into large intact carcasses that would otherwise be unavailable to vultures and, as such, their presence at large carrion benefits the later-arriving vultures. In North America, where Golden Eagles and Bald Eagles often scavenge carcasses, both species dominate vultures and California Condors, although the extent to which such interactions affect the latter remains little-studied. Bald Eagles also routinely chase fully cropped Turkey Vultures from carrion, forcing them to disgorge their meals, and, on at least one occasion kill a vulture that was unable to leave the carcass quickly.

In both North and South America, adult but typically not juvenile, Northern and Southern Crested Caracaras dominate and routinely rob food from feeding Turkey Vultures. Caracaras also dominate Black Vulture at carrion, at least during one-on-one encounters. On the Falkland Islands, groups of Striated Caracaras dominate Turkey Vultures at carcasses, although the reverse often occurs during one-on-one encounters. In Africa, Hooded Vultures interact with both Hooded Crows and

Marabou Storks at garbage dumps and, especially, abattoirs, where they dominate the smaller crows and are dominated by the far-larger storks. The extent to which dominance among these species affects each other ecologically remains unstudied.

SYNTHESIS AND CONCLUSIONS

1. Carrion is neither as scarce nor as unpredictable as many scientists suggest, and vultures are well adapted to secure it.
2. Exclusive carrion diets have both costs and benefits. Carcasses are digestible and nutritious, but many are potentially infectious, toxic, or both.
3. Agricultural practices serve to increase the foraging efficiencies of vultures by concentrating nutritional resources in space and time.
4. Many vultures are direct or indirect human commensals that benefit and are benefited by living in human-dominated landscapes.
5. Vultures require and use low-cost, non-flapping, soaring flight to fly long distances when engaged in low-cost searching flight.
6. All vultures use sight while searching for carrion. A few species also use smell and possibly sound to locate carrion and other nutritional resources.
7. Social facilitation resulting from both single and multi-species aerial information networks makes it possible for vultures to follow one another to carrion, which is an essential component of effective food finding in vultures.
8. Social facilitation also results in large assemblages of hungry vultures at carrion, which typically creates considerable intra- and interspecies competition.
9. Behavioral competition among vultures at feeding sites is both kinetic and complex, and aggressive interactions can be intense.
10. Larger species behaviorally dominate smaller species at feeding sites, but conspecific groups or gangs of smaller vultures can outcompete larger species.
11. Vultures are sometimes, but not typically, cannibalistic.
12. Vultures sometimes approach and hasten the demise of a dying animal, which they then feed upon even while it is still alive.

5 MOVEMENT BEHAVIOR

> Despite its long history, the study of animal movement has generally fallen toward the margins of ecological research because data gathered from wild animals were too sparse to accurately describe these phenomena.
> **Roland Kays et al. (2015)**

DAILY, SEASONAL, ANNUAL, and multi-annual long-distance movements provide vultures with the ability to monitor changes in their nutritional resources and to adjust their breeding seasons and geographic distributions accordingly. Enhanced by soaring flight, low-cost short- and long-distance movements also allow vultures to locate mates and avoid potential predators. Not particularly migratory as a group, vultures include several long-distance migrants, and movements within home ranges of many sedentary species rank among the most expansive of any terrestrial vertebrates.

Often perceived as slow flyers because of non-flapping flight, the speeds at which vultures travel across landscapes rival those of many avian species, including those of many predatory raptors. Vultures are also some of the highest flying of all birds. Although not known for extensive overwater travel, several species breed on remote islands in both the New and Old World, and many occur as vagrants on others. In reality, vultures do get around, and their movements, seasonal, age-specific, and otherwise, substantially affect their distributions.

Recent improvements in satellite bio-tracking, including the development of small, lightweight, and largely aerodynamically neutral signaling devices, enable researchers to extend the distances at which they can detect and follow vultures, and vulture biologists are now tracking tagged individuals across enormous multinational home ranges, as well as along transcontinental migration corridors. Satellite tracking, which has been in widespread use for several decades, allows scientists to work in a golden age of movement ecology, during which they are rapidly developing a more robust understanding of the extent to which avian scavengers travel through their ecological neighborhoods.

TYPES OF MOVEMENT

Dispersal and the geographical population shifts associated with it are fundamental traits of many organisms. Both wide-ranging and largely sedentary animals disperse geographically over time, as do plants. Many life forms, including plants and marine invertebrates—think dandelions and oysters—do so passively in atmospheric, freshwater, and marine currents. Others do so actively, but only briefly, during seasonal reproductive phases of their life cycles, whereas still others, including many vertebrates, such as vultures, do so across more protected periods that can extend for months or years. Most scientists distinguish between *natal* versus *breeding* dispersal, with the former occurring when juveniles or subadults move away from their birthplace before breeding, and the latter when adults move from their birthplace to new breeding sites. Although the term *dispersal* has several evolutionary connotations, I use it here to describe the movements of recently fledged vultures from areas of high adult population densities to areas of lower adult populations at the end of post-fledging parental care, as well as movements that occur among adults selecting breeding areas.

Ranging movements, which consist of movements within home ranges, the areas that individuals routinely visit to obtain the resources needed to survive and reproduce, are typically but not always local. In many vertebrates, home ranges are small enough to defend from nonfamilial conspecifics, and as such are territorial areas of exclusive use. Other, more widely ranging vertebrates, including marine mammals and pelagic seabirds, have considerably larger and not necessarily exclusive areas. The home ranges of vultures typically fall into the latter. Here I describe and detail the various home ranges of vultures and condors, how they differ among species, and how those in many species, especially those of migrants, vary seasonally.

Migration, the large-scale seasonal shifts in populations between breeding grounds and winter areas, which is a signature trait in many birds, is far less common in vultures than in other avian groups. Obligate avian scavengers migrate mainly along north–south axes, or between mountains and less mountainous areas. Seasonal changes in both food and atmospheric energy drive the seasonal migratory movements of vultures, and their dependence on soaring flight for long-distance movement typically limits their movements to terrestrial routes with few, if any, water crossings. The seasonal movements involved can range to several thousands of kilometers. In the Old World, most recently fledged Egyptian Vultures move from Europe to the Arabian Peninsula and northern Africa in boreal winter. In the New World, substantial portions of the North American breeding populations of Turkey Vultures move from North America into Central America and South America in boreal winter.

Dispersal

The onset of natal dispersal varies enormously among vultures. In some species, especially those with protracted breeding seasons, dispersal is initiated when parents react aggressively to their young, jabbing at and preventing them from landing at natal nests when they return to beg for food. This behavior typically occurs as parental individuals ready themselves for the following breeding season. Fledglings in migratory populations, including those of young Egyptian Vultures, Griffon Vultures, and Cinereous Vultures, appear to disperse earlier than those in sedentary populations.

Although we know little of events leading to the onset of natal dispersal in most species, its significance is substantial. The degree to which young disperse prior to breeding has important ramifications in species distributions as well as in geographic differences in local and regional abundances.

The American ecologist Joseph Grinnell suggested that "accidentals" or "vagrants" sighted outside the normal ranges of a species were not accidental at all, but rather were part of routine evolutionary exercises that enabled species to track geographical shifts in potentially suitable landscapes. To make his point, in 1922 Grinnell suggested that it was only a matter of time until the bird list of California would be identical with that for the whole of North America because of accumulated vagrants. Grinnell referred to so-called vagrants as pioneers that were being spun out of growing populations elsewhere. In fact, the known ranges of several species of vultures have expanded considerably recently as a result of such vagrancy. New World Turkey Vultures and Black Vultures have increased their ranges substantially with the former now breeding as far north as western Canada, and the latter now breeding in the northeastern United States and southeastern Canada, as well as in many parts of central and southern South America. Their expansion has been occurring on the heels of expanding human populations. In the Old World, Griffon Vultures have done so in northwestern Africa, as growing numbers of young from the successful Iberian population have dispersed south across the Strait of Gibraltar. Although expansions into previously unoccupied areas are more obvious than are shifts in population densities within existing species ranges, the latter are also likely occur as has happened with Cape Vultures, where individuals have repopulated historically extinct colony sites.

In spite of advancements in satellite tracking, natal dispersal remains little studied in vultures, in part because it occurs during a relatively brief period in their life histories in a time of particularly high mortality. For many young, natal dispersal is best viewed as a necessary evil. Because of high post-fledging mortality during this period, many scientists avoid satellite tagging nestlings and recently fledged young because of the high risk of mortality and the likelihood of abbreviated tracking data. But fortunately, several studies have been conducted involving recently fledged tagged birds and they provide important insights regarding movement ecology during this critical life stage. One such study involved 69 recently fledged wing-tagged Cinereous Vultures in East Gobi Province, Mongolia, near the northern limits of the species range, and reported 21 mainly winter re-sightings of nine wing-tagged individuals in Russia, China, and, mainly, South Korea. The shortest minimal distance traveled by individuals in the study was that of an emaciated individual recovered in eastern China, 650 mi (1080 km) from its tagging site. The longest was that of a bird initially re-sighted wintering in South Korea, and then summering in Siberian Russia. Six other tagged young also overwintered in South Korea, confirming an east–west migration flyway for dispersing young in the species. A second study, involving movements of six nestling Cinereous Vultures tracked for up to 42 months from nesting sites in the Caucasus mountains of Georgia and Armenia, determined that the birds had traveled 150 to 3200 mi (250–5300 km) southeast into Azerbaijan, Iran, and the Arabian Peninsula, after having remained close to their natal areas for several months post fledging. Five had spent at least one summer or winter, or both, in Azerbaijan, while summering in close proximity and wintering farther apart, while taking brief episodic movements of up to 950 mi (1600 km). Intriguingly, none of the birds had established home ranges near their natal sites at the end of the study. Overall, such studies suggest considerable natal dispersal in young Cinereous Vultures, presumably while sampling feeding areas and looking for mates and potential nest sites. All but two of 14 Cape Vultures banded as nestlings at a colony site in the Magaliesberg Mountains of South Africa returned to breed in the region, one at its natal colony site and another at a colony site 15 mi (25 km) from its natal site, suggesting that this species, like colonial-nesting Griffon Vultures in France, demonstrate a degree of natal philopatry. In Griffon Vultures, philopatry appeared dependent on the geography of exiting populations, with young released from captive breeding sites typically selecting the closest nearby populations for breeding from a matrix of traditional regional colony sites.

Four of 66 Egyptian Vultures leg-banded as nestlings in a migratory population in northern Spain also returned and nested close to their birth sites (i.e., 5–30 mi [8–48 km]), although one later nested in France 330 mi (550 km) northeast of where it hatched. Captively bred Bearded Vultures released in the Alps also showed natal philopatry, with maximal settling distances of 30 mi (49 km) from their release sites. Similar evidence of philopatry has also been reported for young Bearded Vultures released in the Pyrenees, where dispersal distances averaged 28 mi (47 km). The degree to which young are able to nest close to their natal sites is likely influenced by the presence of already existing nesters, including parental birds, in territorial species.

The most complete study of natal dispersal in vultures involves American ecologist Patricia Parker's studies of the individually marked population of Black Vultures in central North Carolina described in Chapter 2, in which movements of more than 340 wing-tagged individuals, including six radio-tagged birds were, in some cases, followed for several years. Based on repeated observations of marked individuals, Parker reported that recently fledged young maintained close contact year-round while engaging in nearly daily mutual allopreening, feeding, and flight intercessions. Parker also found that young remained in close and almost daily contact with adults at both communal roosts and feeding sites for at least 6–8 months to several years, and that adults routinely fed their young for as many as 8 months after fledging, with a few parents doing so for several years. Parker noted that closely related individuals, including recently fledged young, were more likely to roost together in the same communal roost on specific evenings.

Ranging Movements

The rate at which vultures discover and consume depends heavily on their searching needs and abilities. Although aerial information networks described in Chapter 4 contribute substantially to this, so does the effective searching behavior of individual birds, a phenomenon that has been difficult to assess before satellite tracking. When used in conjunction with wing tagging and careful behavioral observations, satellite tracking vulture movements over lengthy periods provides a critical framework for understanding how scavenging raptors scour the landscapes they inhabit for largely ephemeral, nutritional resources. A recent summary of the results of 39 such studies provides details into the complex nature of vulture ranging behavior. Published in 2019 by a team of French and Israeli movement ecologists led by Olivier Duriez, the work includes information involving 9 of 16 Old World and 4 of 7 New World species, offering a useful overview of findings to date.

Despite the tropical nature of many vultures, studies involving tropical populations are relatively few. Much of what we know regarding their ranging behavior involves social species in temperate mountains and arid savannas. As might be expected, territorial species including Lappet-faced Vultures and Bearded Vultures apparently forage over smaller areas than do non-territorial species including White-backed Vultures. Daily distances traveled calculated for the 13 species of vultures and condors varied from a low of 3 mi (5 km) in recently fledged Cinereous Vultures to more than 100 mi (160 km) in Griffon Vultures. Values such as these typically reflect the timing between locational fixes, and studies limited to those with about one-hour intervals between locational fixes suggested daily movements averaging 21 mi (35 km), and ranging from 18–36 mi (30–60 km), with considerable geographic differences among several species, along with sizeable overlap across species including Turkey Vultures, White-backed Vultures, and Griffon Vultures. Home ranges, too, vary considerably, depending both on technical analyses and on the species involved, with commensals demonstrating smaller home ranges than non-commensals. In migratory species such as Egyptian Vultures and

Turkey Vultures, temperate zone breeding ranges tend to be smaller than tropical overwintering areas, presumably because of central-place foraging associated with brooding eggs and feeding nestlings. Breeders also tend to have smaller home ranges than non-breeders and juveniles.

An expansive satellite-tracking study undertaken by South African vulture specialist Lindy Thompson and collaborators that followed the movements of Hooded Vultures caught and tracked from The Gambia, Ethiopia, Kenya, Botswana, South Africa, and Mozambique illustrates the flexible nature of within-species differences in the size of home ranges, both within and among geographic populations, as well as annual differences within individual birds. Thompson and her collaborators tracked 30 individual vultures of various ages by satellite for 604 "bird months." Monthly home ranges expressed as 95% KDEs averaged 1420 mi^2 (3735 km^2) and 4845 mi^2 (12,500 km^2) in the largely non-commensal *pileatus* subspecies, that were 31 and 103 times larger, respectively, than the commensal *monachus* subspecies in western Africa, where home ranges averaged 47 mi^2 (121 km^2). For the first 4 months out of the nest, age had little effect on the sizes of mean home ranges. However, immediately post-fledgling three South African birds tagged as nestlings exhibited 4-month post-fledging cumulative home ranges averaging 340 mi^2 (880 km^2). By contrast though, sex had a large effect on the sizes of monthly home ranges. Specifically, the home ranges were significantly higher in males than in females, even though sizes during regional breeding versus non-breeding seasons differed only slightly. Individuals in Ethiopia and The Gambia, where the species is commensal, were recorded in urban areas 11–30% and 10–20% of the time, respectively, whereas largely non-commensal individuals in Kenya, Botswana, South Africa, and Mozambique were reported in urban areas <0.1% of the time. Thompson's results clearly demonstrate that human commensalism has a major influence on the geography of ranging behavior in this species, with commensals exhibiting far less movement in human-dominated landscapes with, presumably, more predictable nutritional resources as compared to non-commensals in more natural areas.

The extent to which commensalism affects reproductive success in the species remains unknown, and although there are no indications of clutch sizes in the typically one-egg clutch nesting species differing regionally, the highest densities of Hooded Vultures anywhere occur in West Africa. Thompson's results also suggest that despite apparent biparental care, females demonstrate considerably lower movement levels geographically, most likely in response to a greater role in both nestling care and nest protection in a species that in South Africa must protect its nest from potentially usurping conspecifics and species, year-round. A greater message from this study—the likes of which have yet to be undertaken in other species—is that movement ecology expressed as home-range size sometimes varies considerably both among regional populations and sex, and studies that attempt to assess such behavior should take this into account.

Part of the challenge in drawing wide-ranging conclusions about the searching behavior of vultures lies in two areas, one methodological and the other ecological. Scientists typically design their studies, including their methods, to effectively assess specific questions about a species without considering how their specific approaches may affect comparing their findings with those of other studies of other species. Although doing so enables science to move forward quickly on a species-by-species basis, it also limits our ability to highlight similarities and differences among species. An even larger difficulty in drawing wide-ranging conclusions lies in the enormous range of ecological circumstances that confront vultures searching for nutritional resources, as well as in the many ways that large-scale cross-country soaring flight allows them to do so. These concerns suggest that scientists will be studying vulturine searching behavior for some time. Standardizing statistical methods used to describe this behavior, together with focusing on and measuring "daily maximal displacement" across species, layering on-the-ground observations with tracking information, and increasing the use of large-scale collaborative studies will allow them to do such studies faster and more efficiently.

Migratory Movements

Although vultures are not as migratory as are predatory raptors, many vulture species engage in routine seasonal movements and several do so over long distances. As in other birds, vulture migration is more common in temperate areas than in the tropics. Notable temperate zone breeders include Turkey Vultures and Black Vultures in the New World and Egyptian Vultures, Cinereous Vultures, and Griffon Vultures in the Old World. Turkey Vultures, Egyptian Vultures, and Cinereous Vultures all routinely engage in one-way migrations of more than 600 mi (1000 km). Several tropical populations of Turkey Vultures and Black Vultures also migrate, most likely to avoid competition with the seasonal arrival of migratory North American populations of Turkey Vultures. Several species, including Egyptian Vultures, Bearded Vultures, Himalayan Vultures, Indian Vultures, Red-headed Vultures, and Rüppell's Vultures, migrate short distances altitudinally. Others, including White-rumped Vultures, Indian Vultures, Griffon Vultures, Lappet-faced Vultures, and Hooded Vultures, migrate short distances in response to seasonal rains.

In North America, where the magnitude of raptor migration in autumn is routinely monitored at more than 130 locations, many raptor migration watch sites do not include observations of migrating Turkey Vultures and Black Vultures in their counts because observers claim that they cannot separate migratory from non-migratory, local movements. And in Central America, early observations of seasonal movements of flocks of Black Vultures were similarly dismissed, despite considerable evidence to the contrary. In both Africa and southern Asia, seasonal movements often occur across broad fronts rather than along narrow migration corridors where movements are more conspicuous, as well as within the year-round distributional ranges of the species involved, both of which make vultures less likely to be considered migratory. Species specifics follow.

Griffon Vultures. Thousands of juvenile and subadult, together with smaller numbers of adult individuals, routinely circumvent the Mediterranean Sea while traveling to overwintering areas in northern Africa. Individuals crossing at the Strait of Gibraltar in southern Spain and Gibraltar typically number between 3000 and 4000 annually. Several dozen individuals also pass into Africa via the Suez, and at least a handful do so at the wider Sicilian Straits in the central Mediterranean and at Bab al Mandeb at the southern end of the Red Sea. Observations at the Strait of Gibraltar indicate that flight behavior is significantly impeded by the 9 mi wide (14 km) sea channel, with many birds attempting passage for weeks before crossing, and others not crossing at all and overwintering in southern Spain. Water crossings at the Strait are largely restricted to between 11:00 and 14:00 on days with light or variable winds and on strong winds from the northwest (i.e., times when being blown out into the Atlantic Ocean is least likely). Crossings are not attempted on days with strong winds from the south or east. Vultures at the site cross in flocks of 20 to 60 individuals, and few, if any, individuals attempt to do so alone. Field evidence indicates that many individuals soar during crossing attempts, at least until they fly out of sight at approximately 2.5 mi (4 km) south of southern Spain. Furthermore, southbound individuals flap more than 10 times as frequently over water as when over land, typically flapping intermittently in episodic flapping bouts of about 5 flaps. Those that flap more than 20 times per minute usually abort crossings and return to Spain in flapping and descending gliding flight, suggesting that over-water passage at the Strait is limited by a physiological or anatomical inability to flap continually.

Egyptian Vultures. As recently as the early 1990s, as many as 10,000 Egyptian Vultures were estimated to migrate from breeding areas in Europe and western Asia to overwintering areas in northern Africa. Although numbers are lower now, most appear to enter Africa via the Strait of Gibraltar in the western Mediterranean, the Strait of Sicily in the central Mediterranean, the Gulf of Suez at the northern end of the Red Sea, and the strait of Bab al Mandeb at the southern end of the Red Sea. By far, the most significant passage occurs at Gibraltar, where at least 3500 southbound migrants cross each autumn, and in northern Egypt, where up to several thousand individuals cross in autumn. As in Griffon Vultures, most water crossings are made in the middle of the day on light tailwinds if available, typically in small groups of several to several dozen individuals. Many first-year and older subadults oversummer in Africa, presumably while fine-tuning foraging skills. Adults precede younger birds, particularly on return migration. In Israel, migrants perform an elliptical migration with many returning individuals following a more westerly route in spring as a result of being deflected northward by the Gulf of Suez, as compared with the autumn when birds follow more easterly routes.

The movements of both adult and first-year Egyptian Vultures have been tracked by satellite. One study involving the 2010–2014 autumn migrations of 19 juveniles from a declining Balkan population indicated that nine individuals died during their first southbound migration while attempting to cross the eastern Mediterranean rather than circumvent it en route to likely overwintering areas in northeastern Africa, perhaps because they had not followed more experienced, older individuals around the water barrier. Another study tracked the migrations of two juveniles tagged as nestlings in France, and a third juvenile tagged as a nestling in Bulgaria. Both French birds began migrating in late August and crossed Gibraltar en route to Africa in mid-September. The Bulgarian bird began migrating in late August to early September and crossed Suez in September. All three individuals migrated at least 2100 mi (3500 km) over the course of 20–25 days, averaging 45–75 mi (77–125 km) a day. The French birds overwintered in the Sahel of southern Mauritania, with one returning to northern Morocco, presumably en route to France when it was 3 years old. The Bulgarian bird spent most of its time in southeastern Chad in the Sudano-Sahelian zone, but it also briefly reached the Central African Republic. Winter ranges of the French birds averaged about 17 mi^2 (45 km^2) MCP, suggesting habitual use of human trash sites.

Two Egyptian Vultures tagged as adults in Spain overwintered in southern Mauritania, northern Senegal, and northern Mali, with one individual returning to the same wintering rounds in a second year of tracking. Both birds performed elliptical migrations in Africa, traveling inland in autumn, and coastally and more westerly in spring. Maximum and average daily travel distances ranged from 240–317 mi (398–529 km) in autumn, and 206–414 mi (443–690 km) in spring, and 147–203 mi (245–339 km) in autumn and 89–157 mi (149–262 km) in spring, respectively. Recent tracks of additional adults largely confirm these findings.

A large collaborative data set involving 188 satellite-tracked migrations of 94 individuals tracked across 70% of the species range detected high migratory connectivity (i.e., different geographic populations overwintered in different geographic areas), relatively diffuse connectivity within regional subpopulations, and wintering ranges up to 2500 mi (4000 km) apart in individuals breeding in the same areas. Individuals breeding in the Balkans and Caucasus traveled faster, up to twice as far, and took longer to complete their journeys than those from Western Europe, most likely because of the greater number of water barriers on the eastern flyway. Despite considerable weather-related differences in travel speeds and the durations of individual migrants among years, individuals exhibited significant overwintering site fidelity. Overall, spring movements were more protracted than autumn movements range-wide.

Cinereous Vultures. Altitudinal migration occurs in many regions, and longitudinal movements of adults between Mongolia and South Korea appear to be increasing, most likely because of the widespread and increasing number of vulture restaurants in South Korea.

Black Vultures. Although more sedentary than Turkey Vultures, many temperate and at least a few tropical populations migrate in response to seasonal shifts in the availability of nutritional resources. Individuals overwintering near the northern limits of their range in eastern North

America migrate coastally and south in response to lower winter temperatures, snow cover, or both. In eastern Pennsylvania, USA, individuals at year-round communal roosts shift daily searching efforts in winter, especially during harsh weather. By contrast, at least some populations in eastern North American undertake reverse migrations, flying northward in early winter to feed on deer offal left by hunters.

Roadside surveys in Panama and the llanos of north-central Venezuela indicate that whereas Black Vultures generally outnumber Turkey Vultures in northern summer, the reverse is true in winter. Two researchers independently reported autumn migration of Black Vultures in Costa Rica and Panama as early as the 1960s, but others questioned whether these observations reflected migration or were simply local movements. That thinking changed, however, when more recent studies confirmed that the arrival of migratory and larger-bodied *meridionalis* subspecies of Turkey Vultures disrupted the feeding behavior of both the local *ruficollis* subspecies and the local *brasiliensis* subspecies of Black Vultures to the extent that both often migrated to avoid the long-distance migrants.

Turkey Vultures. North American populations of Turkey Vultures are the most migratory of all vultures, and their movements are the best studied. Although many populations are non-migratory, northern and central North American breeders migrate both short and long distances, as do many South American individuals from the temperate southern cone of that continent. In eastern North America, the *septentrionalis* subspecies tends to be less migratory than the western *meridionalis* populations, many of which are trans-equatorial migrants that overwinter as far south as Colombia and Venezuela. Argentinian populations in Patagonia also routinely migrate, sometimes trans-equatorially, with at least a few traveling as far north as Colombia in austral winter.

Overall, both the geography and timing of North American populations are better known than those of Central and South American populations, both because of a greater network of raptor migration count sites in Canada and the United States than are found farther south, and the fact that the movements of many of North American breeders have been satellite tracked. Indeed, it is fair to say that we know almost as much about the migratory movements of North American Turkey Vultures as we do about those of any migratory raptors, excepting those of Ospreys and Peregrine Falcons.

Most eastern Canadian and northeastern-most United States breeders evacuate their breeding areas beginning as early as late August, with many staging and feeding farther south in the northeastern United States for a month or more before overwintering in the southeastern United States. At the Hawk Mountain Sanctuary count site in eastern Pennsylvania, peak numbers of migrants are seen in October and early November. In many areas in eastern North America, numbers at roosts grow as migrants join residents and search for carrion with them while amassing fat to help fuel subsequent migrations.

In southeastern British Columbia, western Canada, *meridionalis* migration peaks in early September, and shortly thereafter in western Oregon and southern California, USA, with many passing by the millions in October and November in coastal Veracruz, Mexico, and shortly thereafter, the Canal Zone of central Panama, where in 2016 two million migrants were counted on a single day. In easternmost Panama, the flight apparently fans out quickly as the flight enters northwestern South America, with many individuals continuing southeast of the Andes, and others continuing southeasterly into western and central Venezuela.

Autumn flights north of Mexico typically start in midmorning and peak in early afternoon. However, when the flight reaches coastal Costa Rica, large numbers of migrants are aloft at dawn, and the flight extends well into late afternoon, with early morning migrants soaring in weak, shallow, coastal-water thermals that form as cooler nighttime air is heated and rises above the warmer coastal waters.

Field evidence indicates that little, if any, feeding occurs during migration south of the southern United States, where enormous flocking effectively precludes it. Migrants are considerably lighter upon reaching northern Venezuela, suggesting that fat accumulated before and early during migration fuels much of their low-cost soaring flight. Individuals feed ravenously after reaching northern South America, where an ongoing dry season increases carcass availability, allowing migrants to regain body mass prior to northbound movements the following spring.

The less well-documented return movements are noticeably smaller, presumably the result of overwinter mortality as there are no reports of delayed return migration on the part of subadult birds. Most individuals reach North America in March and April, with populations near the northern limits of their range following the melting snow line north, which exposes large numbers of previously snow-covered carcasses.

Both migrants and non-migrants have been tracked by satellite since the mid-1990s, with more than 100 individuals tagged in eastern and western North America and in the southern cone of South America, one for more than a decade. Results suggest substantial variation in the timing, speed, and geography of movements, both within and across populations, with long-distance migrants traveling faster overall and being less likely to feed en route than short-distance migrants, and with most exhibiting strong migration connectivity, including both breeding and overwintering site fidelity. (See Chapter 2, Box 2.1 for additional details.)

American researcher Somayeh Dodge and colleagues analyzed the migratory movements of 24 Turkey Vultures from four geographically distinct populations representing three subspecies from areas in both North and South America to reveal the extent of regional variation in the movements of these subspecies. Individuals followed for as many as 8 years included breeders from West Coast California, East Coast Pennsylvania, and Midwestern Saskatchewan, as well as from southern-cone Argentina. They demonstrated significant variation in both flight speed and straightness indexes across seasons, both within and among geographic populations. Daily movements were faster and more directional during both outbound and return migrations than were daily movements during non-migratory periods.

East Coast birds, which migrated the shortest distances of the populations studied, traveled at 65–85% slower speeds than did vultures in the other three populations. Both thermal uplifts and ambient temperatures positively affected flight speed during migration, but landscape **N**ormalized **D**ifferences **V**egetation **I**ndexes (NDVI, an index of vegetation productivity) did not. Long-distance, trans-equatorial migrants from the interior of North America traveled faster when migrating across higher latitudes on both their outbound and return migrations than they did at lower latitudes, as well as faster than did short-distance migrants. In all four populations, migrants traveled faster and engaged in migratory flights longer each day on return migrations than on outbound journeys, most likely because of stronger thermals in spring than in autumn. Finally, time taken to complete outbound versus return migrations differed relatively little among the four populations.

A key prediction in understanding why some species of vultures are widespread and abundant, whereas others are not, is that the former are more broadly tolerant of, and adaptively responsive to, environmental conditions, than are the latter. Variability documented in the movement ecology of Turkey Vultures supports this prediction. Overall, recent findings indicate that the species' key innovation of the dihedral-wing posture stabilizes flight in boundary-layer turbulence, which both enhances and enables low-cost movement in a variety of circumstances that allow Turkey Vultures to modify movements to suit local situations better than if the movements were more expensive metabolically. Although the extent that such variability occurs at the individual level remains unclear, the degree of flexibility in vulture migration behavior—both within and among populations—suggests that evolution has selected for different movement strategies in different populations as well as for increased phenotypic plasticity in the species.

Finally, even a data set as robust as the one presented above for Turkey Vultures leaves many questions unanswered, illustrating the need to move beyond accumulating large numbers of small data sets and analyzing them in isolation. Researchers will need to find collaborators with whom to coordinate data collection and will also need to become more comfortable with the idea of data sharing. Sophisticated bio-tracking and the emerging conceptional framework of movement ecology now provide researchers with the opportunity to maximize the benefits of deploying tracking devices to test predictions regarding the whys and wherefores of vulture movements, migratory and otherwise.

SYNTHESIS AND CONCLUSIONS

1. Short- and long-distance movements via low-cost, soaring flight allow vultures to expand their ecological neighborhoods considerably and to better assess and monitor daily the multi-annual changes in essential resources, as well as to fine-tune their breeding seasons and geographical distributions.
2. Natal dispersal to so-called nursery areas allows recently fledged vultures the chance to develop the skills needed for successful reproduction and long-term survival.
3. Territorial vultures tend to have smaller home ranges compared to non-territorial vultures.
4. Home ranges are smaller in tropical than in temperate zone populations, and are smaller in adults than in younger individuals.

5. Home ranges vary both within and among geographic populations, sometimes by more than an order of magnitude.
6. Although not recognized as being exceptionally migratory, many populations of vultures do migrate, both as adults and when young.
7. Long-distance migratory species include Mongolian populations of Cinereous Vultures overwintering in peninsular South Korea; European and Middle Eastern populations of Egyptian Vultures overwintering in the Arabian Peninsula and northern Africa; and both northern and southern Temperate Zone populations of Turkey Vultures, overwintering in the Neotropics.
8. Because many migrations occur within the geographic distributions of the species involved, the migratory behavior of several species remains cryptic and under-reported in the literature.
9. Tropical populations of Turkey Vultures and Black Vultures engage in reciprocal migrations in which individuals migrate in response to the arrival of competitively superior temperate zone migrants.
10. Satellite tracking is providing remarkable new insights into the migratory and other movements of vultures and is vastly improving our understanding of their movement ecology.

6 SOCIAL BEHAVIOR

> Vultures are excellent subjects for a study of social behavior because they regularly live in groups.
> **Jerome McGahan (1972)**

BECAUSE OF THEIR often-communal habits, obligate scavenging raptors rank high as intriguing vertebrates for scientists interested in understanding the causes and consequences of social behavior. Regrettably, this facet of their behavior has not been particularly well studied overall.

Broadly defined, social behavior consists of interactions between and among conspecifics that benefit one or more of the individuals involved; a symbiosis of sorts that occurs mainly within, but also often among, species. Well-structured in insects—think of the caste systems of bees, ants, and termites—social behavior occurs frequently in birds, sometimes even among the pre-hatched embryos of precocious young whose inter-egg vocalizations help synchronize their hatching. Common in many species, most classifications of avian "sociality" are based on adult spacing behavior or on mating systems.

With a few notable exceptions, birds of prey are considered to be primarily non-social, particularly when searching for prey. Vultures are quite the opposite. Most species spend significant time in groups, both when searching for and when feeding upon carcasses and other nutritional resources. Many species roost communally, and several species breed colonially as well. Although vulture social behavior has been studied in some detail, the extent to which it shapes the distributions and abundances of scavenging raptors has yet to receive careful attention.

The reason for this is simple enough. Like field biologists everywhere, students of vulture biology are taught that sample size matters, and that anecdotal information, although often intriguing, is of little value unless bolstered by reams of numerical data. And furthermore, successful contributions to the scientific literature typically require statistical analyses of repeated observations. Studying how social behavior varies both within and among species requires a broader approach than simply studying it in high-population-density areas, and we remain a long way from being able to do that. We also are far from understanding the role of social behavior in affecting the distribution and abundance of vultures.

All 23 of the world's vultures are considered social, but some species are far more so than others. Larger vultures are frequently characterized as being eagle-like, especially during breeding, when nesting pairs defend territories, or areas of exclusive use, from other members of their species, as well as during feeding, when they tend to be predatory and largely solitary, both when searching for and when consuming prey. These species also are less likely to roost communally than are smaller species. Larger species sometimes breed in loose colonies, and massive Griffon Vultures and Cape Vultures routinely nest in colonies. By comparison, smaller species typically feed and roost together, sometimes by the hundreds.

One of the more striking and well-documented interspecies comparisons of social behavior involves the world's most abundant and widespread vultures, the Black Vulture and the Turkey Vulture, which, although ecologically similar in many ways, differ considerably in their social behavior.

Both species have enormous distributional ranges (8 million mi^2 [>20 million km^2]) that stretch across more than 100 degrees of latitude from Canada in northern North America to Tierra del Fuego in southernmost South America. The two also are similar phylogenetically, with fossils of both dating from the mid-Pleistocene and splitting less than 5 million years ago. Both species nest similarly, although Turkey Vultures nest at slightly greater nearest-neighbor distances. Both roost communally, often comingling when they do. Both are currently expanding their distributional ranges in the Americas, and both are increasing numerically in many parts of their ranges. The two, however, differ behaviorally, anatomically, and physiologically in many ways, which together explain notable differences in their social behavior.

Turkey Vultures, which fly with wings held in a strong dihedral, are able to hug surrounding landscape in low-altitude, soaring flight while searching by smell and by sight for both small and large carcasses. Black Vultures typically fly higher on horizontally held wings in soaring flight similar to that of *Gyps* vultures. Turkey Vultures, together with their tropical congeners, the Lesser Yellow-headed Vulture and the Greater Yellow-headed Vulture, have a keen sense of smell that enables them to detect carcasses by smell as well as by sight. Black Vultures, which cannot smell carcasses directly, often do so "parasitically" by following Turkey Vultures to carcasses the latter have located.

A second distinguishing feature is an enhanced physiological ability to endure water stress in Turkey Vultures, which is lacking in Black Vultures. Although both are more abundant in the tropics than in the temperate zone, Turkey Vultures are notably more common in deserts than are Black Vultures. Experimental and field evidence suggest that the Turkey Vultures are better equipped than Black Vultures to conserve the "metabolic water" they create when processing their food, a physiological distinction that became apparent to me several years ago while studying the movement ecology of the two species in the Sonoran Desert of south-central Arizona, USA. The work involved trapping and wing-tagging Black Vultures to study their short-distance, local movements, and trapping and satellite-tagging Turkey Vultures to study their longer-distance migrations. Trapping occurred at sites baited with carcasses of stillborn calves provided by local dairy farms and with drinking water offered up in a shallow basin. Although both species were attracted to, and fed upon the carcasses, only Black Vultures drank from water basins at the sites, both before and after having fed. Although it did not occur to me at the time, the difference in drinking behavior helps explain substantial differences in the regional distribution and abundances of the two species. Extensive route surveys of the two species within and outside of the tropics indicate that the highest densities of Turkey Vultures anywhere occur in coastal and near-inland Chile in natural areas along the edge of the Atacama Desert. Black Vultures rarely, if ever, resided in such locations, except in urban areas and other human-dominated landscapes. Similarly, the northern limits of breeding ranges of Turkey Vultures extend well into Canada and throughout most of the arid sections of the American West, whereas those of the Black Vultures largely stop before reaching the deserts of the American West, except in human-dominated areas where surface water is available. How the physiological difference in water metabolism came about remains unstudied, but the Black Vulture's greater dependence on sources of open water may have fostered its development as a frequent inhabitant of human-dominated landscapes where open water is almost always available.

Another significant difference between the two species involves communal roosting. Throughout temperate zone North America, where the roosting behavior of both species is well studied, Turkey Vultures and Black Vultures often comingle at communal roosts. Satellite-tracked Turkey Vultures, however, do not necessarily roost at a single site, but often move among roosts up to 31 mi (50 km) apart, whereas Black Vultures that are wing tagged as family groups often roost together at single sites for years.

The extent to which the differences mentioned above affect differences in social behavior has yet to be studied—as does the idiosyncratic nature of social behavior in most other species.

DEVELOPMENT OF SOCIAL BEHAVIOR

Several core aspects of vertebrate social behavior, including species recognition, are under so-called phase-sensitive, or specifically timed, developmental control. Indeed, both filial and sexual recognition, or the abilities to recognize members of the one's species and appropriate species-specific mates, respectively, often are learned by individuals during brief critical periods early in life: first, when nestlings fixate their attention on the first object with which they have visual, auditory, or tactile experiences; and later in life, when older young fixate on conspecifics. This so-called imprinting is both rapid and limited to brief critical periods that are independent of external stimuli. Other aspects of socialization, including interaction with mates regarding parental roles such as incubation and nest defense and interaction with conspecifics and members of other species while feeding, are decidedly more nuanced and individualized. Those aspects develop experientially during protracted periods of up to several years.

Although recently fledged, less social feeders, such as Turkey Vultures and Bearded Vultures, receive episodic instruction from their parents post-fledging while feeding with their parents, many aggressive group-feeding species, including Old World *Gyps* species, apparently do not receive instruction. Black Vultures are exceptional in that recently fledged and relatively naïve young remain in family groups for 6 months or more, during which time older family members both experientially teach and protect young while their offspring learn how best to feed successfully in aggressive groups with non-relatives. In many other species, including most if not all *Gyps*, recently fledged young expand their home ranges considerably at about the time that parents begin taking care of their next brood. As a result, many first-year young spend up to several years in what are called nursery areas where in competition with similarly aged, naïve young, individuals build skills that allow them to compete with and successfully become breeding adults. Although such behavior has yet to be studied, pre-breeding subadults from phylogenetically distinct species of facultative avian scavengers exhibit behavioral similarities that apparently occur in at least some species of vultures as well. The young of both Northern Ravens and Striated Caracaras, two phylogenetically and somewhat ecologically similar species of facultative scavengers, feed opportunistically on the carcasses of both dead and dying vertebrates in gangs of as many as several dozen or more individuals that often overwhelm and outcompete single or paired adults at carcasses in much the same way groups of Old World *Gyps* and New World Black Vultures outcompete adults of both their own and other, sometimes larger, species at carcasses. This social behavior, which in many ways resembles that of human teenage street gangs, appears to be of particular significance in the young of at least some non-vulturine avian scavengers. How it functions in young vultures has yet to be explored.

SYNTHESIS AND CONCLUSIONS

1. Vultures are considered social species in part because they often roost communally and feed in groups.
2. One of two of the world's most common and widespread obligate avian scavengers, the Turkey Vulture, is relatively non-social, whereas the other, the Black Vulture, is highly social.
3. Several key innovations, including an acute sense of smell and the ability to soar effectively at low altitudes, most likely enable Turkey Vultures to be less social than most other species.
4. Communal roosts function more as short-term motels than as long-term homes for Turkey Vultures, but the reverse is true for Black Vultures.
5. Age-specific gang behavior is a useful behavioral adaption for some species, and a lifelong behavioral adaptation for others.

6. Overall social behavior is best thought of as a behavioral tool for obligate scavenging birds of prey, the functionality of which varies considerably among species.
7. Despite its likely significance in vulture biology, the ontogeny and maintenance of social behavior in many species has yet to be studied in detail.

7 VULTURES AND PEOPLE

Vultures have a tough time in the public eye. . . . 75% of people think of [them] as dirty undertakers.
BirdLife International (2020)

RELATIONSHIPS BETWEEN HUMANS and vultures date back nearly 2 million years to a time when ancestral protohumans were following soaring vultures to carcasses on the savannas of East Africa. More recently, pastoralists and transhumant populations turned the table on this symbiotic relationship as our ancestors concentrated large herds of domesticated ungulates that vultures followed and fed upon as predictable carrion resources. Modern hunter-gatherers, including the Hadza of northern Tanzania, continue this relationship today, and although populations of several vultures now function as human commensals in both suburban and urban areas of the Americas, Africa, and southern Asia, many human interactions currently threaten and endanger their regional and at times global populations (Table 7.1).

In part because they are both large and conspicuous, vultures have long histories in numerous human cultures, including frequent references of exposing human carcasses to them for consumption. Artifacts from the Neolithic Turkish town of Çatal Hüyük suggest that such practices were underway at least as far back as 7000 BC. More recently, ancient Egyptian goddesses Nekhbet and Mut were associated both as hieroglyphic Griffon Vultures and Lappet-faced Vultures, with the former as the powerful patron of Upper Egypt, and the latter as a nursing mother of the dead, as well as the womb from which new life emerged. Vultures also figured conspicuously in Egyptian art, including on the death mask of Tutankhamun and a spread-winged necklace from his tomb. The Egyptian Vulture, referred to as pharaoh's chicken and sacred to the Egyptian god Isis, was once protected by law and often mummified.

Table 7.1. Conservation Status of the World's Vultures and Condors

Species	Current global conservation		
	Distribution[a]	Status (and date first listed[b])	Human threats
New World Vultures			
Black Vulture	NA and SA	LC	
Turkey Vulture	NA and SA	LC	
Lesser Yellow-headed Vulture	MA and SA	LC	
Greater Yellow-headed Vulture	SA	LC	
King Vulture	MA and SA	LC	
California Condor	NA	CR (1988)	Lead poisoning; Habitat degradation
Andean Condor	SA	NT (2000)	Direct persecution; Poisoning
Old World Vultures			
Palm-nut Vulture	AF	LC	
Bearded Vulture	AF, EU, OR	NT (1988)	Direct persecution
Egyptian Vulture	EU, AF	EN (2007)	Veterinary drug poisoning; Direct persecution
Hooded Vulture	AF	EN (2007)	Direct persecution; Bushmeat
Indian Vulture	OR	CR (2000)	Veterinary drug poisoning
Slender-billed Vulture	OR	CR (2000)	Veterinary drug poisoning
White-rumped Vulture	OR	CR (2000)	Veterinary drug poisoning
Griffon Vulture	EU, OR, AF	LC	
Rüppell's Vulture	AF	CR (2007)	Poisoning; Habitat degradation; Direct persecution
Himalayan Vulture	OR	NT (2014)	Veterinary drug poisoning
White-backed Vulture	AF	EN (2007)	Poisoning; Habitat degradation; Direct persecution
Cape Vulture	AF	EN (1988)	Poisoning; Habitat degradation; Direct persecution
Red-headed Vulture	OR	CR (2004)	Veterinary drug poisoning; Land-use change
White-headed Vulture	AF	CR (2007)	Habitat degradation; Poisoning
Cinereous Vulture	EU, OR, AF	NT (1998)	Persecution; Habitat degradation
Lappet-faced Vulture	AF	EN (2000)	Poisoning

Source: BirdLife Red List as of March 2020 (www.birdlife.org/datazone).
[a] AF = Africa, AU = Australasian Region, EU = Eurasia, MA = Middle America; NA = North America, OR = Oriental Region, SA = South America.
[b] LC = Least Concern, NT = Near Threatened, CR = Critically Endangered, EN = Endangered, Threatened [obsolete], VU = Vulnerable.

In the New World, cultural relationships with vultures and condors date from prehistoric times. The remains of California Condors were buried in California shell middens, and pictographs of the species appear in the state's San Rafael Mountains. Condor feathers were used in religious and ceremonial practices in many coastal cultures in the North American

Pacific Coast, and their wing bones, like those of Neanderthal-era Griffon Vultures, were used as instrumental flutes.

With the exception of several conspicuous, large-bodied vultures, including California Condors, Cape Vultures, Bearded Vultures, and Cinereous Vultures, avian scavenging birds of prey were rarely mentioned by conservationists before the 1980s. Even the iconic and critically endangered California Condor received only three pages of discussion in James Greenway's *Extinct and Vanishing Birds of the World* in 1958; one half-page in Guy Mountfort's *Rare Birds of the World* in 1988 (where it is mentioned somewhat disparagingly, "No bird in the world has had more money or expertise lavished on it"); and as little more than a "Pleistocene relic" in Dominic Couzens' *Atlas of Rare Birds* in 2010.

In fact, most obligate scavenging raptors were largely inconspicuous on the radar screens of global avian conservation until the 1990s, when two ornithologists surveying vulture populations in Africa and southern Asia brought their conservation status to the attention of the conservation community. The first was the French raptor specialist Jean-Marc Thiollay, who had previously surveyed vultures along thousands of roadways and tracks in West and Central Africa in the 1970s. The second was Vibhu Prakash, an Indian scientist who had earlier conducted a PhD study of raptors inhabiting the 11 mi^2 (29 km^2) Keoladeo National Park in Rajasthan, north-central India in the 1980s. Thiollay and Prakash reached similar conclusions. Both noted that populations of once common *Gyps* vultures had declined precipitously in both Africa and southern Asia by more than 90% in less than several decades, suggesting catastrophic population collapses in many formerly widespread species. Although neither Thiollay nor Prakash documented conclusively the underlying causes for the declines, both stated that numerical collapses had indeed happened, and that the declines had occurred in alarmingly brief periods of time.

The response from the avian conservation community was immediate. A Google search for the phrase "vulture conservation" that had yielded but 38 publications on the subject before 2000, yielded 228 publications in 2000–2009, and 719 in 2010–2019, suggesting that vulture conservation had finally come of age. Although single-species concerns, including those involving Andean Condors, California Condors, Cape Vultures, Bearded Vultures, and Cinereous Vultures, very much remained in play, multi-species, trophic-level concerns focused on entire communities of vultures have grown dramatically.

PRINCIPLES OF VULTURE ENDANGERMENT

Working with input from BirdLife International and others, the International Union for the Conservation of Nature (IUCN) Red List of threatened species actively inventories and assesses the conservation status of birds, globally. In 1999, the IUCN Red List ranked only four of the world's 23 species of vultures and condors, or 17% of all obligate scavenging birds of prey, as Threatened, with most endangerment caused by habitat loss and human persecution. Twenty years later, in 2019, IUCN ranked 16 species, or fully 70% of the world's scavenging raptors, as Near Threatened (4), Endangered (5), or Critically Endangered (7) (see Table 7.1).

Although human activities threaten species in dissimilar ways, several overarching principles of endangerment apply to this trophic assemblage. Understanding these principles allows one to anticipate future endangerments and to act to prevent them. Below I highlight each in turn.

Principle 1: There are three general categories of human threats to vultures, and their effects can be additive.

These categories include 1) land-use change including the loss of natural areas; 2) misuse of environmental contaminants, including pesticides; and 3) full-frontal assault, including purposeful poisoning, shooting, trapping, and eating.

Threats often work additively. Land-use change, for example, can force vultures and condors into suboptimal agricultural areas where both intentional and unintentional poisoning and direct persecution can occur, and the misuse of environmental contaminants, including poisons used to control mammalian pests and predators, can kill vultures unintentionally.

Principle 2: Land-use change remains the most important threat to many vultures.

This principle is true particularly when associated with large-scale agriculture and increases in human density. Although several species of vultures, including human commensals such as Black Vultures in Central and South America, and Hooded Vultures in West Africa, thrive in human-dominated landscapes, many others do not.

Principle 3: Intrinsically rare vultures can be at greater risk of endangerment and extinction than common and widespread species; however, even common species can be endangered.

Species such as California Condors, Himalayan Vultures, and Bearded Vultures, which depend heavily on mountain updrafts for soaring flight, and, as such, occur at low densities in relatively uncommon and often low-productivity mountainous regions, have a greater risk of endangerment because their small populations provide little buffer against rapid environmental change. Common species can be endangered when threats include highly toxic environmental contaminants or widespread persecution.

Principle 4: Obligate scavenging raptors are at considerably greater risk than are predatory raptors.

Several biological adaptations of vultures render them particularly vulnerable to a variety of human threats including poisoning, direct persecution, and habitat loss. Poisoning is of great concern because vultures feed on dead animals that are sometimes poisoned; they frequently eat large amounts of food at single meals and thus are more likely to consume fatal doses of environmental contaminants; and they search for and feed together in groups at so-called common tables that expose large numbers of individuals to potentially fatal toxins simultaneously. Because many large vultures evolved in conditions that selected for low annual mortality, as well as their having some of the lowest reproductive rates of any birds—with some species laying single-egg clutches, or foregoing annual reproduction entirely—and because vultures often require high population densities for information transfer in locating carcasses, population declines resulting from human threats are likely to worsen as social networks collapse.

HISTORICAL HUMAN THREATS TO VULTURES AND CONDORS

Direct Persecution

The history of human persecution of vultures differs considerably from that of human persecution of predatory raptors. Whereas targeted shooting and trapping has long affected the latter, prior to modern sanitation, vultures were often valued and protected by humans. For example, Black Vultures were welcome human commensals and mutualists in the Americas throughout much of the 19th century, and today, the same is true for Hooded Vultures in rural, suburban, and urban areas in West Africa. Human attitudes toward vultures, however, quickly shifted as human sanitation improved. Because vultures are not charismatic, large numbers feeding voraciously on domestic carcasses in human landscapes are often unjustly accused of killing livestock.

The mistaken notion that vultures spread, rather than help contain, human and livestock diseases such as anthrax, botulism, and hog cholera also leads to population control, particularly when local numbers increase. This misperception was certainly true for Black Vultures in early and mid-19th century southeastern North America, which was a time when potentially vulnerable livestock were allegedly being preyed upon by Black Vultures whose populations were increasing rapidly. In the early 1950s, the American wildlife biologist Paul W. Parmalee studied the species in eastern Texas, USA, to assess local movements. This study was at a time when local landowners were actively controlling Black Vultures, which were believed to "attack and devour domestic animals, particularly the young" in large numbers. Parmalee trapped and banded more than 450 individuals to determine the size of their home ranges in response to such complaints. He sent questionnaires to landowners to learn more about their control efforts, which, in some instances, dated back to the late 1800s. Local farmers quickly pointed out that although Turkey Vultures sometimes joined Black Vultures in feeding on "ill-gotten" carcasses, the former hardly ever took part in killing live prey. Although neither species was then protected by law in Texas, more than 80% of the landowners indicated that their real target was the more aggressive Black Vultures, which they box-trapped, killed, and removed in large numbers. One farmer reported catching 200 vultures during his first trapping effort, along with an average of 150 individuals monthly thereafter.

More than 20,000 Black Vultures had been banded in the United States by 1952, almost entirely in the Southeast. Parmalee's analysis revealed that almost all of the 899 vultures that had subsequently been recovered had either been trapped or mysteriously found dead. Many of the farmers Parmalee queried had initially kept records of numbers killed, but they stopped counting after the first 500 or 1000 caught birds, with most suggesting that trapping and killing vultures was successful in reducing the loss of livestock, especially during lambing and calving. Parmalee estimated that at least 100,000 birds had been trapped and killed, and that this figure was "without question conservative," as many farmers queried had not responded. Although both Turkey Vultures and Black Vultures remained protected in most of the United States, Parmalee re-trapped only 6% of the more than 400 he had banded, a percentage he attributed to a high rate of unreported persecution. He concluded that trapping and killing constituted the principal means of Black Vulture population control in Texas for more than a half century.

Texas was not alone in its efforts to control Black Vultures. In mid-20th-century North America, neither Turkey Vultures nor Black Vultures were protected in Texas, Tennessee, Louisiana, Florida, North Carolina, and South Carolina, where large and growing populations of occasionally predatory Black Vultures occurred. Although both species are currently protected under various laws and treaties in the United States, Canada, and Mexico, in 2020 the United States Fish and Wildlife Service continued to issue so-called allowable takes of problematic Black Vultures from several hundred individuals annually in states near the northern end of the species to more than 250, 000 annually over the entire eastern United States, an area in which growing regional populations were estimated at more than 4 million birds in 2015, and where the allowable take in Virginia, USA, had grown 62% to more than 5700 during the previous decade.

Although direct persecution of New World species is usually limited to Black Vultures, Turkey Vultures, too, have been targeted, particularly on islands, including the Falkland Islands, where the latter has expanded its feeding behavior in the absence of Black Vultures. On the island archipelago the species reportedly prey on newborn lambs, on cast sickened or injured fallen sheep, and on the eggs of Upland Geese; hence, the local government issues depredation permits. By far the most significant example of the Turkey Vulture predation behavior occurs on islands along the western coast of South America that are free of Black Vultures. The oceanic Humboldt Current upwelling of nutrient-rich cold, deep water and associated super-abundant anchovies attract hundreds of thousands of breeding and guano-producing seabirds to an area where thousands of egg- and nestling-eating Turkey Vultures together with hundreds of predatory Andean Condors once gathered. Large-scale shooting of both species peaked in the early 1900s (Box 7.1).

Box 7.1 Protecting Guano Birds from Andean Condors and Turkey Vultures

The most striking example of large-scale direct persecution of Turkey Vultures and Andean Condors occurred in the early 1900s when a Peruvian government agency known as *Compañia Adminstradora del Guano* (CAG) hired dozens of sharpshooters to kill hundreds, if not thousands, of both species that were feeding on the eggs, nestlings, and fledglings of "guano birds," which included several species of colonial-nesting seabirds that were then considered by many to be the most valuable birds in the world.

Between 1840 and 1880, Peru exported more than 12 million tons of guano (the Spanish word for the Native American Quechua word *wanu*), the excrement of seabirds and bats, for use as an exceptionally effective nitrogen, phosphorus, and potassium-rich fertilizer during a so-called guano rush. More than 50 million seabirds, including, most notably, hyper-abundant Guanay Cormorants, Peruvian Boobies, and Peruvian Pelicans that nested on islands off coastal Peru in non-El Niño years when coastal upwelling associated with the Humboldt Current provided them with abundant nutrient resources. (El Niños bring warm surface waters from the Pacific Ocean, which prevents the upwelling of the colder, nutrient-rich water.) The colonies annually produced tons of economically valuable, nitrogen-rich excrement that, over the course of many thousands of years, had accumulated guano to depths of more than 160 ft (50 m) at many seabird colony breeding sites. Numerous predatory gulls and skuas, together with Andean Condors and Turkey Vultures, preyed on the eggs and chicks of the guano birds, and all were targeted and shot on sight as part of the government's comprehensive conservation policy.

The extent of incessant institutionalized shooting was considerable. During three years in the late 1930s, records indicate that the government agency bought 47,500 shotgun cartridges to kill the birds. Although the large-scale killing of condors at first led the American ornithologist Robert Cushman Murphy to suggest that it was "criminal that sharpshooters . . . should be employed for the express purpose of killing such magnificent creatures as condors." However, after witnessing 18 condors methodically eliminating the reproductive output of one island's entire breeding colony over the course of several days as individual birds "walked from one burrow to another, as if to sit, ogre-like, and await the exit of an unsuspecting victim," Murphy remarked that his "sympathies inclined more toward the victims" as well as toward the actions of *CAG* in removing the "veritable harpies."

The subsequent development of a commercial fishery focused on anchovies—the principal food of guano birds—for use in the production of high-protein livestock food, coupled with the occurrence of several upwelling-less El Niño years in the 1950s, led to significant declines in seabird populations, which, subsequently, led to the cessation of large-scale vulture and condor persecution the following decade. Today, oceanic upwelling from the Humboldt Current continues to support populations of more than 4 million seabirds together with substantial numbers of southern sea lions and their carcasses, which result in some of the highest regional densities of Turkey Vultures anywhere, including populations that extend south well along the coastline of northern Chile where communal roosts associated with coastal municipal dumps routinely host more than 400 birds.

Populations of European vultures, too, have been labeled as predatory, and consequently, have been persecuted well into the 21st century, particularly in areas where large numbers of hawks and eagles are shot in large numbers. Egyptian Vultures disappeared from the Alps near Geneva between 1910 and 1915 during a period of nest plundering and shooting. Bearded Vultures, too, were persecuted throughout the 19th century in the Piedmont Alps, where one of the most common hunting techniques involved placing a carcass in a ditch and beating to death the scavenging raptors that landed to feed upon it. More sophisticated control efforts included shotguns, with individual birds being taken for taxidermy as well as for supposed game management. Remarkably, persecution of reintroduced Bearded Vultures continued in the Alps into the early 1990s, before finally subsiding at the turn of the 20th century as regional breeding populations took hold. In northern Italy, Griffon Vultures were shot while migrating to wintering areas in the Balkans in 1930s and 1940s, as well as near breeding sites in Sardinia in the 1960s and 1970s. In 2007, individuals of an increasing Iberian population of Griffon Vultures were shot in response to official carcass-removal programs that were put in place because of the spread of the livestock disease Bovine Spongiform

Encephalopathy (BSE). The birds involved, which had been feeding on cattle and other livestock carcasses placed in shallow pits known as *muladares*, had difficulty locating sufficient food and began attacking vulnerable livestock. Feeding stations for vultures have since mitigated the problem, and human persecution of the Griffon Vultures has lessened.

Bushmeat and Witchcraft

Although not as significant as in predatory raptors, the African bushmeat trade, along with that of body parts used in witchcraft and traditional medicine, remains a regional threat in West and Central Africa, where the carcasses of both shot and poisoned vultures are openly sold in bazaars. A recent study of the trade in several dozen markets in 12 West African countries found that common species of vultures were sold more frequently than less common species, with an annual take of Palm-nut Vultures of 975–1460 individuals amounting to 0.7–1% of the regional population. The use of vultures in witchcraft and traditional medicine varies greatly among and within African countries, with the dried and pulverized brains of Hooded Vultures sold as a useful clairvoyant for predicting the outcomes of sporting events. Although human consumption of vultures appears to be unusual outside of Africa, reports suggest that late 20th-century consumption of vulture carcasses in southeastern India also was considerable.

Environmental Contaminants

Organochlorine pesticides. Developed to replace considerably more toxic inorganic pesticides, synthetic organochlorine pesticides (OCPs) were used in World War II to protect combatants from insect vectors of human diseases. Widespread global use of OCPs for managing both human and agricultural pests came about shortly thereafter. The most notorious OCP is DDT (**d**ichloro-**d**iphenol-**t**richloroethane), an inexpensive, broad-spectrum, and persistent pesticide that was first used against insect pests in Europe and North America in the late 1940s and early 1950s, and worldwide shortly thereafter. Far less toxic to vertebrates than the inorganic first-generation pesticides it replaced, "wonder pesticide" DDT earned the Swiss entomologist Paul Müller a Nobel Prize in physiology and medicine in 1948.

American ornithologists began to notice declines in reproductive success in many species of predatory birds, including Bald Eagles, as early as the 1940s, as well as in Peregrine Falcons shortly thereafter. However, it was not until early 1965 that a group of concerned North American scientists met at the University of Wisconsin to discuss the pesticide situation. Increased eggshell breakage in Peregrine Falcons suggested a physiological connection. Several years later, British falcon specialist Derek Radcliffe published data that tied the timing of eggshell thinning in his study site to the widespread use of DDT. With science on the case, things began to happen quickly.

Two years after Radcliffe's analysis, controlled experiments in the United States involving captive American Kestrels conclusively demonstrated evidence that DDT disrupted the ability of many birds to produce sufficient eggshell material. Armed with this new information, conservationists successfully initiated bans on the widespread use of DDT and other organochlorine pesticides across most of North America and Western Europe in the early 1970s, which quickly led to reductions in contaminant levels in many areas and a reversal in eggshell thinning. Shortly thereafter, raptor populations that had suffered through DDT-era lows began to rebound. In several cases, natural increases were assisted by reintroduction efforts, and by the mid-1980s, widespread recoveries were well underway. In North America, Black Vultures, Turkey Vultures, and California Condors all exhibited eggshell thinning during the DDT era, and populations of the first two declined in many areas in the 1950s through the 1970s before reversing pre-DDT-era declines that had been underway. The use of OCPs in Mexico, which continued into the 1980s, was also linked to eggshell thinning in local populations of Black Vultures. Data indicated considerable eggshell thinning in the eggs of California Condors during the DDT era, and that hatching success was lower in a reintroduced population that had particularly thin eggs. Incubating parents appeared to crack eggs prior to the use of DDT, and in 1989 the American raptor specialist Lloyd Kiff suggested that it "seems reasonable to suspect that severe eggshell thinning experienced by condors during the DDT era further exacerbated their tendency to break eggs, resulting in an increased rate of egg failure."

In Africa, where many persistent pesticides have yet to be banned, there is less evidence of OCP impact. Eggshell thickness indexes collected for Cape Vultures and White-backed Vultures in southern Africa before and during the DDT era were 3.5% and 3.7%, respectively, which were well below the critical level for increased eggshell breakage. Blood and tissue levels of organochlorine pesticides collected in the early 1990s in Africa from White-backed Vultures, Cape Vultures, and Lappet-faced Vultures were low compared with results documented for other species, suggesting that populations from which the samples obtained were under no direct threat from OCPs. Nevertheless, persistent OCPs continue to potentially threaten migratory populations of vultures when encountered in less-developed nations in the tropics and subtropics of Africa and South America where populations overwinter close to or in agricultural areas.

More recently developed pesticides. Newer, third-generation pesticides that replaced persistent, second-generation OCPs brought both good and bad news. The replacements include the organophosphates malathion and parathion, and the carbamates carbofuran and aldicarb. Although far less persistent than their OCP predecessors in having half-lives of from several hours to several days, they kill pests by inhibiting cholinesterase, an enzyme that turns off the neurotransmitter acetylcholine in the nervous systems of both insects and vertebrates, and, therefore, is acutely toxic to both groups, including vultures and humans. Unfortunately, examples are numerous of this more recent pesticide threat, including the widespread use of carbofuran as an inexpensive and readily available poison in predator control.

Lead. The stable, corrosion-resistant, soft, and malleable heavy metal, lead, which is easily worked by humans, has been in human use for at least 6000 years. Early Egyptians used it for figurines, and Romans expanded its use in plumbing. Unfortunately, lead mimics and replaces calcium in many vertebrate organs and as such is highly toxic in both humans and birds. At low doses, lead can affect many aspects of avian behavior, immunity, and reproductive success. Lead poisoning in waterfowl, which sometimes ingest lead shot as grit for the mechanical digestion of plant materials, was first recognized in 1919. Its use in ammunition during waterfowl hunting eventually was banned in the United States and Canada in the 1990s, and in many countries shortly thereafter.

Both vultures and condors are highly susceptible to lead poisoning when they ingest lead shot and bullet fragments in the carcasses shot by hunters using lead ammunition. Lead toxicity may have played an important role in the decline of the California Condors as early as the 1800s, and today it remains a major barrier to the successful recovery of the species even though its use in ammunition for most big-game hunting has been banned in California since 2008. Recently, several populations of free-ranging wild California Condors were routinely re-trapped, and those whose blood lead levels were above the clinical normal range received chelation therapy prior to their release. This medical procedure involves the administration of chelating agents to remove heavy metals from the body. Turkey Vultures collected in coastal Monterey County, California, USA, prior to regulations on the use of lead ammunition and then recaptured a year later, had a significant twofold decrease in blood lead levels following a ban on lead, suggesting that additional regulation of its use remains merited for this species. Although information is scarce regarding lead exposure in Andean Condors, a study involving 152 feather samples collected in 15 communal roosts throughout most of the species' Patagonian range suggested low concentrations from ammunition used in big-game and hare hunting. A more recent study documented exposure throughout the species range in Argentina, as well as numerous lead exposure events in Chile, Ecuador, and Peru.

In the Old World, Egyptian Vultures on the Canary Islands and mainland Iberia exhibited lead contamination in both blood and bone tissue. Concentrations in the blood were higher during hunting seasons; concentrations in the bone tissue increased with age of the bird; and the degree of bone mineralization was lowest in individuals with the highest bone lead concentrations. Insular individuals had lower reproductive success and higher bone tissue concentrations overall than did mainland individuals, and it seems reasonable to assume that lead toxicity occurs when female vultures mobilize bone tissue during egg laying and transmit lead accumulated in bone to their embryos.

All but 2 of 23 Griffon Vultures trapped in southern Spain in 2003 had measurable levels of lead in their blood, and as recently as 2016 multiple cases of lead poisoning were confirmed in Spanish populations. Lead exposure in the Bearded Vulture and White-backed Vulture in southern Africa also has been reported.

The degree to which lead exposure affects vultures other than California Condors is not well studied. Six captive Turkey Vultures orally dosed with BB-sized lead shot for 6 months varied in their responses to dosing with large numbers of shot (3 or 10 per bird). One high-dose bird died within 143 days, and one low-dose bird survived the 211-day length of the trial, much longer than reported for other species of birds tested, including California Condors and Bald Eagles. Unfortunately, responses of similarly dosed Andean Condors appear to mimic California Condors, indicating that vultures vary in their response to this toxic heavy metal.

Veterinary drugs. Unfortunately, pesticides are not the only agricultural chemicals threatening vultures. Indeed, the largest vulture die-off attributed to environmental contaminants involved a seemingly innocuous veterinary drug, diclofenac. The story of diclofenac's impact is of special note because it involved population crashes of three formerly common species, the White-rumped Vulture, the Indian Vulture, and the Slender-billed Vulture. In the mid-20th century in India, White-rumped Vultures once nested in densities of approximately 3 pairs per km^2 (or 9 pairs per square mile) within metropolitan Delhi. They had a global range that once included most of the Indian subcontinent as well as much of Southeast Asia, and they once ranked as the most abundant large raptor on earth. Between 1980 and 2000, however, this species had all but disappeared in the wild, along with both the Indian Vulture and the Slender-billed Vulture (see Table 7.1). In fact, numbers of the two most abundant species, the White-rumped Vulture and the Indian Vulture, had declined by more than 95% in just two decades, during which numbers of the decidedly less common Slender-billed Vulture had declined by more than 90%.

Box 7.2 Uncovering Diclofenac

The most catastrophic large-scale vulture kill attributed to environmental contaminants did not involve pesticides, but rather an otherwise seemingly benign veterinary drug. The story begins on the Indian subcontinent, where in the mid-1990s the Indian ornithologist Vibhu Prakash first noted dramatic declines in populations of several species of common and widespread vultures in a former vulture stronghold, Keoladeo National Park in Rajasthan, India. The birds in question included three species of *Gyps* vultures, five of which occur in India.

The simultaneous and catastrophic declines of three populations of many once common and widespread vultures in southern Asia in the late 1980s and 1990s were alarming for several reasons. First, with minimum estimated declines of 96% for White-rumped Vultures and 92% for Indian Vultures and Slender-billed Vultures, the rapidity with which the three species disappeared was unprecedented. By comparison, although regional populations of the cosmopolitan Peregrine Falcon had undergone similar declines in some areas during the DDT era, the global fate of that species was never in doubt. This was not the case for the three species of vultures, all of which were in precipitous decline throughout their ranges in southern Asia. Second, and perhaps more disturbing, was that no one was able to explain why the populations were declining. The usual suspects, habitat loss, pesticides, and direct persecution, offered little in the way of answers.

Although scientists had known for some time that the increased and largely uncontrolled hunting of wild ungulates in Southeast Asia had decimated the food base for vultures there, large numbers of livestock carcasses remained readily available in India and Pakistan, and vulture populations had crashed there as well. Whereas some pointed to direct persecution as the cause—an argument intuitively supported by the region's high human densities—the vultures in question were declining throughout their ranges, particularly in India where all three were valued for their role as environmental scavengers and, in certain parts of society, held in cultural and religious esteem. At first, pesticide or heavy-metal poisoning was thought likely, but exhaustive toxicological analyses of carcasses failed to detect clinical levels of these environmental contaminants. Given these dead ends, the conservation community began to focus on the possibility that an unknown and particularly virulent pathogen was killing the birds.

The argument for a novel infectious disease to which the vultures were immunologically naïve made sense. After all, many of the affected birds looked sick. Neck drooping and lethargy were typical prior to death, and not one declining breeding colony had recovered once high mortality set in. Autopsied carcasses from India and Pakistan revealed renal malfunction and visceral gout in most individuals, along with a high frequency of enteritis, conditions that are often reflective of infectious diseases. As recently as 2003, most conservation scientists believed that an epidemic disease, probably viral, was their most promising lead.

At much the same time, Lindsay Oaks, an American avian pathologist at Washington State University, in Washington, USA, had reached a far different conclusion. Oaks, whose analyses of vulture tissues had eliminated the widespread use of pesticides and heavy metals as the culprit, had turned to analyzing fresh carcasses for signs of fragile viruses and bacteria. Finding none, by late 2002, Oaks had concluded that an epidemic disease was unlikely. Determined to discover an explanation for the ongoing tragedy, Oaks began to focus on a series of less likely contaminants, including several veterinary drugs that had recently come into widespread use in livestock in the region. In early 2003, the new work paid off. Tests revealed that all 25 of the vultures Oaks examined that had died of renal failure had been exposed to the drug diclofenac, a non-steroidal, anti-inflammatory painkiller, or NSAID, that had come into medical use in the United States in the 1980s. Furthermore, none of the 13 vultures that he examined that had succumbed from other causes showed any exposure to the drug. In addition, four non-releasable vultures that had been given single oral doses of diclofenac—either at the normal mammalian dosage (in 2 birds) or at one-tenth of that dosage (in 2 birds)—died of acute renal failure within 2 days. Diclofenac's wholesale generic cost in the developing world was less than 2 US dollars per month for human use and far less for veterinary purposes, and the drug had come into widespread use in cattle in India and Pakistan in the 1990s.

Analyses of carcasses from much larger areas of India and Nepal revealed high residues of the drug in dead vultures there as well. Moreover, a simulation model of vulture demography indicated that diclofenac exposure rates as low 1-in-760 available livestock carcasses could have produced the massive die-off. Armed with this newfound information, avian conservationists began calling for a ban on the drug for veterinary use shortly thereafter.

The precipitous declines led conservationists to call for an immediate ban on the anti-inflammatory veterinary drug diclofenac, which acts to induce acute kidney failure in vultures that had consumed livestock that had been treated with the drug but then died. In 2005, the Indian government enacted a 6-month phasing out of diclofenac for veterinary use, and farmers there have been told to replace diclofenac with available alternatives believed to be less toxic to the birds. Other countries, including Pakistan and Nepal, followed suit (Box 7.2). Even so, declines continued well into the first decade of the 21st century, and in 2015 a three-species Indian population estimate indicated that 12,000 birds were all that remained, with numbers of White-rumped Vultures stabilized at 6000, and numbers of Indian Vultures and Slender-billed Vultures still in decline, with the latter estimated to number little more than 1000 individuals.

Indian populations of Egyptian Vultures and Red-headed Vultures also have disappeared rapidly and markedly, although the onset of their declines had begun after those of the three *Gyps*. In 2003, Indian protected-area populations of Egyptian Vultures were only 32% of what they had been in 2000, and those of Red-headed Vultures had declined 16% over the same period. Many researchers have suggested that these two subordinate species were not exposed to diclofenac at large carcasses until the more dominant *Gyps* populations had declined numerically and access to carcasses increased.

Initial disbelief in the avian conservation community soon gave way to the realization that the declines represented the loss of tens of million vultures, making it, arguably, the most cataclysmic of all recent raptor population declines, including those of the 20th-century DDT era, by more than an order of magnitude. Once diclofenac had been identified as the cause for the decline, the hard sell of attempting to protect the physically and ecologically unattractive scavenging birds of prey, many of which were human commensals unanchored to natural habitats, began. In 2011, a consortium of 14 (by late 2019 the consortium had grown to 24) local, regional, and conservation organizations was established—**S**aving **A**sia's **V**ultures form **E**xtinction (SAVE)—with the common understanding that doing so required identifying, prioritizing, and implementing actions needed to prevent species extinction based on sound scientific grounds. By late 2019, SAVE's conservation focus included 1) banning veterinary use of all drugs known to be toxic to vultures and identifying drugs that are nontoxic; 2) promoting "vulture safe zones" across southern Asia, including transboundary cooperative efforts, where known toxic drugs are not used; 3) developing national reporting systems for vulture deaths; 4) implementing high-profile anti-poisoning campaigns in Southeast Asia and encouraging diclofenac bans there; and 5) maintaining vulture restaurants with toxin-free food when appropriate. These actions, together with ongoing captive-breeding and release programs designed to maintain genetic diversity, should help restore historic communities of southern Asia vultures.

Junk. That anthropogenic debris, including bits and pieces of glass, ceramic, plastic, rubber, metal fragments, and bottle tops, occurs in the diets of adult and nestling vultures is a widespread and problematic phenomenon in vultures globally. Micro-trash, including the remains of plastic bags, was found in 34% of the regurgitation pellets of Black Vultures in northeastern Chiapas, Mexico, and in 58% of regurgitation pellets of Turkey Vultures feeding in and around the capital of (and only town in) the Falkland Islands, as well as in 82% and 18% of 102 Turkey Vulture regurgitation pellets collected in the Atacama Desert of Chile, respectively. Micro-trash, which was first reported at a California Condor nest site in 1922, and in nine of ten breeding caves in 2000, was confirmed as a contributing factor in the deaths of four of six dead nestlings between 2002 and 2005. Micro-trash is often fed to the nestling *Gyps*, including Griffon Vultures in both Israel and Armenia, as well as to White-rumped Vultures in Pakistan and to Cape Vultures in South Africa. Possible reasons for young ingesting junk include its being mistaken for bone fragments by parents of calcium-stressed nestlings, whose parent-regurgitated diets are largely devoid of the calcium needed for bone development, or that it may help grind large chunks of offal in the digestive tract, or provide sufficient mass for peristaltic action during pellet to egestion. Whatever its function or functions, junk ingestion is potentially lethal in vultures, and its availability needs be restricted through targeted garbage removal near nests whenever possible. In addition, small bone fragments should be supplied at feeding stations visited by vultures, especially during the breeding season.

Human Structures

Generation, transmission, and distribution of electrical energy. Human use of electrical energy threatens vultures in several ways. Most electric transmission, including both the long-distance transmission from central power sources to distribution centers, and the short-distance transmission to end users locally, occurs via overhead lines because safer underground cables are more expensive in terms of the cost of initial materials used and the installation costs. Despite ongoing attempts to mitigate the impacts of overhead power lines, transmission lines remain a serious threat to large-bodied, aerial-searching species including vultures, and collisions and electrocutions caused by power lines can impact populations, locally and regionally.

The keen eyesight of vultures would seem to reduce the likelihood of collisions with power lines; however, the visual fields of many species have large blind areas above, below and behind the head. Blind spots coupled with inflight downward head positioning, which enhances their abilities to search surrounding landscapes for carcasses and shields their eyes from solar glare, acts to blind them in the direction of forward flight.

Electrocution of perched individuals on poorly designed power poles and stanchions also poses a significant threat. Dozens of migrating Egyptian Vultures traveling along a 19 mi (31 km) long transmission line in northeastern Sudan have been electrocuted across several decades, with many of their carcasses found at the base of ill-designed transmission poles, and similar power-line related electrocutions occurring elsewhere globally. Electrocution and collision deaths along power lines can be mitigated both by reducing food availability where power lines occur while simultaneously developing foraging areas or supplemental feeding sites away from such sites, and by redesigning and retrofitting power line stanchions to reduce electrocutions.

As the number of wing turbines continues to increase globally, this power source, too, continues to pose a threat to vultures. Although vultures often approach active wind turbines, the improperly installed and inappropriately managed arrays of turbines can kill scavenging raptors. Key factors in reducing the impact of turbines include situating them away from vulture colony sites and migration corridors and bottlenecks. Unfortunately, increased demand for electrical power, coupled with widespread industry deregulation, suggest wind turbines will continue to threaten scavenging raptors for some time.

Water troughs. In arid areas with little surface water, vulture drownings in livestock water troughs can be problematic. Records of 163 separate events in South Africa indicate individual and mass killings of 120 Cape Vultures, 67 White-backed Vultures, and 3 Lappet-faced Vultures, with drownings typically occurring during drinking and bathing after individuals have consumed large meals or strychnine-laced bait, which creates a burning sensation in the digestive tract. Drowning occurs predictably in summer, and the younger, less experienced, recently fledged individuals are particularly vulnerable. Such events are largely preventable. Mitigation includes the use of plastic floats and wooden planks or ladders that birds can grasp to extract themselves, nylon netting to prevent vulture entry, and the provision of less dangerous drinking places nearby.

Aircraft. Many cities and towns place garbage dumps in open areas near commercial and military airports, in part because potential crashes, security concerns, and noise associated with low-flying aircraft make such areas unfit for commercial and residential development. Unfortunately, their frequent juxtaposition with municipal dumps also increases the likelihood of vulture–aircraft collisions, particularly for commensal vultures. Between 1955 and 1999, at least 40 serious vulture–aircraft collisions involving military and 6 civilian aircraft occurred in Asia, Africa, Europe, and North America, killing dozens of White-rumped, Long-billed, Egyptian, White-headed, and Griffon Vultures in the Old World, and a number of Black and Turkey Vultures in the New World. Although collisions, which in India were estimated to have cost 70 million US dollars annually in the 1980s, are not, themselves, likely to cause vulture population declines, mass killings of hundreds of vultures in response to such events are. Food availability near airports needs to be restricted globally.

Unintended and Targeted Poisoning

Accidental poisoning, typically aimed at mammalian predators and other noxious wildlife, has threatened vultures for almost 200 years. Strychnine, a bitter and poisonous chemical produced by Asian vomit trees as a defense against herbivores, came into widespread use as a poison against noxious wildlife in the early 1800s. The toxic alkaloid was isolated in 1818 by French Chemists Bienaimé Caveto and Pierre-Joseph Pelletier, who, two years later isolated the anti-malarial drug quinin. Initially used in small amounts to treat human paralysis, highly toxic strychnine kills by asphyxiation, paralyzing muscles in the respiratory tracts of vertebrates. Strychnine was first marketed by the Hudson Bay Company to poison ship rats, and its use against so-called problematic bears, wolves, and coyotes was widespread in the 1830s. Less than 20 years later, a San Francisco newspaper report listed it as killing several California Condors that had fed upon a bear carcass laced with it. The widespread use of strychnine as a vertebrate toxin in the United States continued well into the 1920s when it was employed against Black Vultures, Turkey Vultures, and other avian scavengers. The most recent California Condor kill attributed to it was at a spiked coyote carcass in California in 1966. Regrettably, the highly toxic chemical, which remains registered in the United States for manual belowground applications against pocket gophers, is readily available there and in many places globally.

Unfortunately, poisoning noxious wildlife has a long history globally that predates the use of strychnine. British authorities established an effective vermin control program in the African Cape Colony in 1656 to control carnivores, and today many toxins continue to threaten vultures on that continent. In Kenya, where the readily available, low-cost carbamate pesticide carbofuran is currently the poison of choice, extensive transect surveys conducted in and around the Maasai Mara ecosystem revealed that significant declines of *Gyps* vultures occurred there between surveys conducted in 1976 and 1988 and again in 2003 through 2005, suggesting that declines were most consistent with predator poisoning that targeted lions and hyenas. Hundreds of vultures are killed unintentionally as bycatch during such events, as well as those poisoned specifically for traditional medicine and witchcraft.

In 2013, a team of three African and Scottish researchers reported that most vulture poisonings in Africa were related to illegal livestock predator control, with vultures as unintended victims. More recently, ivory poachers now poison and kill elephants and contaminate their carcasses specifically to eliminate vultures, whose circle soaring above carcasses often alerts game rangers to poaching activities. Indeed, between 2012 and 2014, 11 poaching incidents in 7 African countries killed at least 155 elephants and 2044 vultures, increasing the rate at which vultures were poisoned over 30 times that recorded in earlier incidental poisonings. Poachers also kill elephants by spiking waterholes, as well as by attracting elephants with laced delicacies including watermelons, oranges, and pumpkins, resulting in many nontarget species being killed, including carcass-scavenging vultures.

Africa is not the only place where vulture poisoning remains problematic. Between 1990 and 2017, 211 poisoning events in Spain killed 294 Egyptian Vultures. Most of the events involved adults during the breeding season, with many being killed after the Bovine Spongiform Encephalopathy (BSE) epidemic at a time when victims of Mad Cow Disease were incinerated and not available to avian scavengers. In 2017 and 2018, 66 Andean Condors were poisoned in Argentina, including 34 at a single carbofuran-laced sheep carcass. In Cambodia, recent declines in White-rumped Vultures, Slender-billed Vultures, and Red-headed Vultures have been linked to carbofuran poisoning including indiscriminate carcass baiting and the lacing of natural forest pools.

Overall, vultures are particularly vulnerable to poisoning because of their tendency to feed communally and to gorge themselves at poison-laced carcasses. Unless or until inexpensive poisons including carbamates and organophosphate agricultural pesticides are removed from veterinary offices, pharmacies, and agricultural supply shops and are subsequently banned, poisoning will remain a significant threat to vultures and condors for the foreseeable future.

Global Climate Change

Global climate change can affect the distributions and abundances of avian scavengers in many ways. In North America, Black Vultures and Turkey Vultures have increased the northern limits of their breeding and wintering ranges in part because of warmer winters and because reduced snowfall exposes many carcasses to scavenging. In South Africa, populations of Bearded Vultures have disappeared from many occupied low-altitude breeding sites, including protected places where global climate change has increased human use.

Although models of global climate change differ considerably in terms of regional details, most scientists agree that global warming will result in wet areas getting wetter and wet seasons staying wet longer, and that drier areas will get drier and dry seasons will stay dry longer. Many believe that protracted droughts, including those that have occurred recently in the American West and in sub-Saharan Africa, are likely to increase in frequency as the world warms, and that decreased productivity there will lead to increased desertification. Although episodic, protracted dry seasons and short-term episodic droughts sometimes increase vulture reproductive success, drier, more desert-like habitats, are likely to support smaller and more scattered populations of avian scavengers than will be found in less-dry savannas and grasslands.

The rapidity with which global climate change occurs will determine the extent to which individual species are able to adapt to such changes. Although regional scenarios remain speculative, the overall impact of a warmer and more seasonal world with fewer grasslands and savannas and more deserts does not bode well for many species of vultures, including many *Gyps* species that depend on grassland and savanna environments.

Supplemental Feeding

Authorities date the intentional feeding of wild birds to the 6th century, when a Scottish monk tamed a pigeon by routinely feeding it. Intentional supplemental feeding for avian conservation dates to the harsh British winter of 1890–1891 when newspapers suggested offering food to backyard birds. Feeding wild birds is now common in many countries, including the United States, where it ranks second to gardening as the most popular of all hobbies. In the mid and late 20th century, auxiliary feeding was critical in successfully restoring young peregrines and other predatory birds of prey during successful reintroduction efforts on the heels of the DDT era in the United States.

Vulture restaurants were first used to feed Bearded Vultures and other obligate scavenging birds of prey in Giant's Castle Game Reserve in the Drakensberg mountains of South Africa in 1967. Shortly thereafter, similar supplemental feeding sites were used in reintroductions of Griffon Vultures in southern France and of Bearded Vultures in the southern Alps. Today vulture restaurants occur in many countries, including Nepal, India, Cambodia, South Africa, Swaziland, Spain, South Korea, and the United States, both for conservation and for ecotourism purposes. In India and elsewhere in southern Asia, supplemental feeding is used to provision uncontaminated food to help reduce diclofenac-related mortality at vulture colonies in "vulture-safe zones" where the veterinary drug diclofenac is no longer in use. In the United States, vulture restaurants provide lead-free food to California Condors and attract individual condors for recapture as part of conservation efforts.

As in all human endeavors, despite good intentions, vulture restaurants can be poorly designed and misused. That said, properly employed supplemental feeding has the potential for being an important tool in practical, on-the-ground conservation, as well as in vulture conservation science (Box 7.3).

Box 7.3. Vulture Restaurants

Introduced to vulture conservation in the late 1960s, "vulture restaurants" are now common in both vulture conservation and ecotourism. Appropriately designed and maintained, vulture restaurants serve many conservation actions. However, inappropriately designed or misused vulture restaurants can be problematic. Their benefits and necessary considerations are listed below.

Benefits

- Making additional food available to birds especially in times of food scarcity.
- Provisioning of bone fragments as a source of calcium can reduce calcium deficiencies, osteoporosis, and other skeletal abnormalities in young and older vultures that feed largely on the offal of large carcasses.
- Provisioning poison-free food including lead-free and diclofenac-free carcasses.
- Attracting vultures to areas where they had previously occurred after threats in those areas have been eliminated.
- Diverting vultures from occasional predation on vulnerable livestock, including newborns.
- Raising awareness among landowners, farmers, and the general public.
- Using them as destinations for eco- and photo-tourism.

Considerations

- Vultures may habituate to a specific place and not continue to forage widely. Regular provisioning of small amounts of food at several fixed sites may benefit adult individuals, whereas random provisioning of large carcasses more likely will benefit sub-dominant immature and juvenile individuals.
- Enthusiasm to maintain a restaurant may decline with time and its maintenance cease.
- Maintaining a food supply can become problematic.
- Supplemental food may attract problematic animals including feral dogs, jackals, monitor lizards, and other mammalian and avian scavengers.
- Vulture restaurants may be viewed by local populations as vectors of contagious diseases.
- Locations close to power lines may increase electrocution and collision risk.
- In impoverished areas, meat may be stolen by local inhabitants.

- Vulture restaurants can serve conservation only if clean food is a limiting factor to the populations in question.
- Implement vulture restaurants only after careful observations of the potential vulture populations have been undertaken and long-term local support for restaurants is available.

FINAL THOUGHTS

The lengthy parade of human threats listed above summarizes the long, and frequently unfortunate, relationship between vultures and people. Obligate scavenging birds of prey continue to be threatened by humans one way or another by land-use change, particularly when increased predator control and modern sanitation reduce or eliminate carcasses from human-dominated landscapes. Direct persecution, albeit generally in decline, will continue in areas so long as vultures and their body parts are held in high esteem as clairvoyants and where their soaring over carcasses is detested by poachers.

Overall, land-use change in general and environmental toxicants in particular remain at the top of the list of human threats to vultures, the latter in ways that would have been unthinkable as recently as several decades ago. Because vultures and condors feed voraciously in large groups at communal tables, environmental toxins and other contaminants will continue to threaten this group, until and unless several things change. First, known toxins must be restricted in use or banned outright. Second, population monitoring must be increased in both common and uncommon species so that the existence of new threats is revealed before catastrophic effects have happened. Third, research focused on understanding the limiting factors that affect vulture populations must be expanded.

The recent explosion in interest in rare vultures, although welcome, should be extended to common and widespread species as well. Lesser-studied populations of Red-headed Vultures, King Vultures, and both Greater Yellow-headed Vultures and Lesser Yellow-headed Vultures, for example, should be investigated in increased detail, and their populations better understood and carefully monitored. Threats to existing populations must be recognized rapidly if these species are to remain common and widespread.

Looking forward, it is easy to be discouraged by the current situation. Signs of hope can be found in many places, however, and I remain guardedly optimistic overall. Free-ranging California Condors now range across larger areas and in greater numbers that at any time in my lifetime, and the notion that this Pleistocene relic is close to extinction is fading. Although lead poisoning remains a threat to California Condors and other avian scavengers, conservationists are working diligently to reduce and potentially eliminate this significant threat. Reestablishing it as a functional force in natural and human-dominated landscapes, however, seems unlikely.

Populations of southern Asian vultures appear to have stabilized recently. Reintroductions of captively reared individuals are well underway. The prognosis for this scavenger assemblage is better now than it has been in decades. By contrast, vulture poisoning in Africa and South America remains problematic, and poisoning, both intentional and accidental, is likely to continue for the foreseeable future. The good news is that toxicants are now on the radar screen of many conservationists, and international collaborations now underway are likely to be sustained.

My own experiences of working with local, regional, and international partners in North and South America, Africa, Europe, and Asia to change existing situations for the better reinforce my commitment to the well-established principle of thinking globally while acting locally.

The vulture conservation community is now producing important new information on these often-resilient species, as well as tweaking management strategies in ways that will help reverse declining populations of rare and uncommon species and stabilize populations of those that remain common and widespread.

Commensal and mutualistic relationships between vultures and humans date back millennia. A recent study of the widespread and abundant Turkey Vulture suggests a significant ecological role in organic waste management for this species in support of human activities. The roles that Turkey Vultures, Black Vultures, Hooded Vultures, Indian Vultures, and White-rumped Vultures and other commensal and mutualistic *Gyps* have played, and continue to play, in helping to clean up human-dominated landscapes need to highlighted, so that vultures can be more fully appreciated as essential ecological entities.

Although humanity's relationships with vultures have created some of the most significant avian conservation concerns anywhere, they have also taught many of us that taking this trophic assemblage for granted is no longer tenable ethically. Protecting vultures as essential recyclers in ecosystems is not only ethically correct but also ecologically smart. Choosing to build on existing and growing bodies of knowledge, and working together to protect vulture populations while they remain common and functional ecological entities, will allow us to do just that.

SYNTHESIS AND CONCLUSIONS

1. Symbiotic relationships between vultures and humans date to when prehistoric hominins followed flights of vultures to large-mammal carcasses in East Africa.
2. Vultures are not charismatic fauna, and they often are mistakenly blamed for killing livestock.
3. Scavenging raptors were largely missing in action in avian conservation until 90% declines in previously abundant populations and widespread *Gyps* vultures occurred in southern Asia in the late 20th century.
4. Several biological adaptations of vultures render them particularly vulnerable to human threats. These include feeding on dead animals that have been poisoned, eating large amounts of food at a single meal, and feeding together at "common tables," all of which expose large numbers of vultures to contaminants simultaneously.
5. Today principal human threats to vultures and condors are land-use change; environmental contaminants, including veterinary drugs, the heavy metal lead, and pesticides; poisoning; and persecution, with the latter including catching and killing vultures as pests, for bushmeat and witchcraft, including as clairvoyants, and as sentinels of illegal poaching.
6. Land-use change that includes reductions in wildlife and their carcasses and modern sanitation affect many populations of vultures.
7. Vultures are often killed at poisoned carcasses set out for mammalian predators such as hyenas, lions, pumas, wild and feral dogs, and jackals.

8. Lead poisons vultures when they consume remains of hunter-shot animals.
9. Vultures are killed when they collide with electric-energy lines and wind turbines, and when they are electrocuted when perched on ill-designed power line posts and stanchions.
10. Creating a better understanding of the ecological functions and needs of vultures, along with monitoring their populations, are essential features of effective vulture conservation.

APPENDIX

Scientific Names of Vultures and Condors, and Other Birds, and Scientific Names of Other Animals Cited in the Text

Common and scientific names are the widely held preferences of the scientific and conservation communities of 2019 and primarily follow those of Gill and Wright's *Birds of the World: Recommended English Names* (2006).

VULTURES

Common name	Scientific name
Andean Condor	*Vultur gryphus*
Bearded Vulture	*Gypaetus barbatus*
Black Vulture	*Coragyps atratus*
California Condor	*Gymnogyps californianus*
Cape Vulture	*Gyps coprotheres*
Cinereous Vulture	*Aegypius monachus*
Egyptian Vulture	*Neophron percnopterus*
Greater Yellow-headed Vulture	*Cathartes urubutinga*
Griffon Vulture	*Gyps fulvus*
Hooded Vulture	*Necrosyrtes monachus*
Himalayan Vulture	*Gyps himalayensis*
Indian Vulture	*Gyps indicus*
King Vulture	*Sarcoramphus papa*
Lappet-faced Vulture	*Aegypius tracheliotos*
Lesser Yellow-headed Vulture	*Cathartes burrovianus*
Turkey Vulture	*Cathartes aura*
Palm-nut Vulture	*Gypohierax angolensis*
Red-headed Vulture	*Sarcogyps calvus*
Rüppell's Vulture	*Gyps rueppellii*
Slender-billed Vulture	*Gyps tenuirostris*
White-backed Vulture	*Gyps africanus*
White-headed Vulture	*Aegypius occipitalis*
White-rumped Vulture	*Gyps bengalensis*

OTHER RAPTORS

African Fish Eagle	*Haliaeetus vocifer*
Bald Eagle	*Haliaeetus leucocephalus*
Bateleur	*Terathopius ecaudatus*
Black Kite	*Milvus migrans*
Bonelli's Eagle	*Hieraaetus fasciatus*
Golden Eagle	*Aquila chrysaetos*
Madagascar Serpent Eagle	*Eutriorchis astur*
Martial Eagle	*Polemaetus bellicosus*
Northern Crested Caracara	*Caracara cheriway*
Red Kite	*Milvus milvus*
Southern Caracara	*Caracara plancus*
Spanish Imperial Eagle	*Aquila adalberti*
Steppe Eagle	*Aquila nipalensis*
Striated Caracara	*Phalcoboenus australis*
Tawny Eagle	*Aquila rapax*
Variable Hawk	*Geranoaetus polyosoma*
Verreaux's Eagle	*Aquila verreauxii*

OTHER BIRDS

Chicken (domestic)	*Gallus gallus domesticus*
Giant petrels	*Macronectes* spp.
Greater Flamingo	*Phoenicopterus roseus*
Guanay Cormorant	*Leucocarbo bougainvillii*
Gulls	*Larus* spp.
Grey Parrot	*Psittacus erithacus*
Herons	*Ardeidae* spp.
Hooded Crow	*Corvus cornix*
Hummingbirds	Trochilidae
Lesser Flamingo	*Phoeniconaias minor*
Marabou Stork	*Leptoptilos crumeniferus*
Northern Raven	*Corvus corax*
Oxpeckers	*Buphagidae* spp.
Peruvian Booby	*Sula variegata*
Peruvian Pelican	*Pelecanus thagus*
Skuas	*Stercorarius* spp.
Sunbirds	*Nectariniidae* spp.
Upland Goose	*Chloephaga picta*
Wandering Albatross	*Diomedia exulans*
White-necked Raven	*Corvus albicollis*
White Pelican	*Pelecanus onocrotalus*

OTHER ANIMALS

African elephant	*Loxodonta africana*
African lion	*Panthera leo*
African painted dog	*Lycaon pictus*
American alligator	*Alligator mississippiensis*
American bison	*Bison bison*
Andean fox	*Lycalopex culpaeus*
Asiatic lion	*Panthera leo*
Baboons	*Papio* spp.

Black-and-white tagu	*Tupiambis merianae*
Black wildebeest	*Connochaetes gnou*
Blesbok	*Damaliscus pygargus*
Blue wildebeest	*Connochaetes taurinus*
Bobcat	*Lynx rufus*
Brazilian lancehead	*Bothrops insularis*
Capybara	*Hydrochoerus hydrochaeri*
Caribou	*Rangifer tranandus*
Chacma baboons	*Papio ursinus*
Civet	*Nandinia binotata*
Cow	*Bos* spp.
Dogs	*Canis* spp.
Dolphins	*Cetacea* spp.
Duikers	Cephalophinae
Eland	*Taurotagus oryx*
Fin whale	*Balaenoptera physalus*
Giraffe	*Giraffa camelopardalis*
Goats (domestic)	*Capra hircus*
Gray wolf	*Canis lupus*
Guanaco	*Lama gunicoe*
Hares	*Lepus* spp.
Horse	*Equus caballus*
Hyenas	*Hyaenidae* spp.
Impala	*Aepyceros melampus*
Jackals	*Canis* spp.
Jaguar	*Panthera onca*
Komodo dragon	*Varanus komodoensis*
Lagomorphs	Lagomorpha
Leopard	*Panthera pardus*
Mongooses	*Herpestidae* spp.
Monitor lizards	*Veranus* spp.
Ostriches	Struhionidae
Pallas cat	*Otocolobus manul*
Pangolins	Philodota
Peruvian anchoveta	*Engraulis ringens*
Pig (domestic)	*Sus scrofa*
Pocket gophers	*Thomomys* spp.
Polecat	*Mustelinae* spp.
Puma	*Puma concolor*
Rabbit	*Ictonyx striatus*
Red deer	*Cervus elaphus*
Red-hartebeest	*Alcelaphus buselaphus*
Reindeer	*Rangifer tarandus*
Sardines	Clupeidae
Sharks	Selachimorpha
Sheep (domestic)	*Ovis aries*
Ship rat	*Rattus rattus*
Southern sea lion	*Otaria flavescens*
Springbok	*Antidorcas marsupialis*
Squirrel	Sciuridae
Steenbok	*Raphicerus campestris*
Striped skunk	*Mephitis mephitis*

Tasmanian wolf	*Thylacinus cynocephalus*
Termites	*Termitoidae* spp.
Terrapins	*Emydidae* spp.
Thomson's gazelle	*Gazella thomsonii*
Tortoises	*Testudinidae* spp.
Water buffalo	*Bubalus bubalis*
Weasels	*Mustela* spp.
White-tailed deer	*Odocoileus virginianus*
Wild boar	*Sus scrofa*
Wildebeest	*Connochaetes* spp.
Wolverine	*Gulo gulo*
Zebras	*Equidae* spp.

GLOSSARY

adaptive radiation. The process by which organisms diversify rapidly from an ancestral species into many new forms.

aerodynamic lift. An upward force on a flying wing created as it deflects and forces air downward.

aggressive mimicry. Mimicry in which a predator mimics a non-predatory species in order to deceive and more easily approach its prey.

allometric scaling. The change in an organism's attributes related to proportional changes in body size.

anadromous. A fish species migrating up rivers from the sea to spawn. The opposite of catadromous.

aquiline. Curving like an eagle's beak or resembling an eagle.

aspect ratio. A measure of the wing defined as the squared wingspan divided by wing area. Raptors with long narrow wings have high aspect ratios.

asynchronous. Not synchronous or simultaneous. In birds, asynchronous hatching of a clutch of eggs means that individual eggs hatch on different days, creating a situation in which earlier hatchlings have a developmental "head start" on later hatchlings.

austral summer. Summertime south of the equator, beginning in December and ending in March. *See also* boreal summer

biogeography. The study of the geographical distributions of species and the historical and biological factors that led to—and continue to—shape them.

biological indicator. A species whose presence in an ecosystem is used to indicate the conservation status of an ecosystem. Biological indicators are more easily measured and monitored surrogates for the ecosystem processes they indicate. Many conservation biologists consider raptors, including vultures and condors to be indicators of ecosystem health.

biomagnification. The concentration of toxins in an organism, including vultures, as a result of their ingesting other plants or animals in which the toxins are less concentrated.

boreal summer. Summertime north of the equator, beginning in June and ending in September. *See also* austral summer

broad-spectrum. Among pesticides, those that kill many kinds of pests. The organochlorine pesticide DDT, which kills many kinds of insects, as well as other invertebrates, is a broad-spectrum pesticide.

BSE. Bovine spongiform encephalopathy (aka mad-cow disease) is a neurodegenerative disease of cattle.

bushmeat. The meat of wild animals, including vultures.

C_3 to C_4 grassland transition. C_3 photosynthesis produces a three-carbon compound via the Calvin cycle, whereas C_4 photosynthesis produces an intermediate four-carbon compound. The C_4 photosynthetic pathway is an expansion of the C_3 pathway that evolved in the Oligocene, 25 to 32 million years ago, and did not become ecologically significant until the Miocene, 6 to 7 million years ago, as an adaptation to high light intensities, high temperatures, and dryness. C_4 plants dominate grassland floras and biomass production in the warmer climates of tropical and subtropical regions.

carbamate. A salt or ester derived from carbamic acid containing the anion NH_2COO, or the group $OOCNH_2$.

carbofuran. A vertebrate-toxic carbamate pesticide. Carbofuran is marketed under the trade names Furadan, by FMC Corporation, and Curaterr 10 GR, by Bayer, among several others. It is used to control insects in a wide variety of field crops, including potatoes, corn, and soybeans. Carbofuran also has contact activity against mammalian predators and other livestock pests.

carrion. Dead, putrefying flesh that serves as food for vultures.

cast sheep. A cast sheep is a sick or injured sheep that has fallen or laid down and cannot upright itself. Once down, gases often begin to build up in the rumen (the large first compartment of the stomach of a ruminant), and the individual can die in a matter of hours.

catabolism. The metabolic breakdown of complex molecules in living organisms into simpler ones, together with the release of energy.

central-place foraging. An evolutionary ecology model for analyzing how an organism maximizes successful foraging while traveling through a patch of resources from a home base to discrete foraging sites rather than traveling at random.

cere. A fleshy and sometimes colorful swelling of featherless skin that surrounds the nares of predatory raptors, vultures, parrots, skuas, and pigeons that protects the nostrils.

chain migration. Occurs when migratory populations that breed at high latitudes migrate approximately the same distance as those that breed at lower latitudes, thereby maintaining their latitudinal relationship between seasons. *See also* leap-frog migration; reciprocal migration

charismatic megafauna. Sometimes called *flagship species*, charismatic megafauna are symbolic species whose conservation status is used by conservationists to capture public attention. The protection of charismatic species can help protect the ecosystems upon which they depend. Large raptors, including some vultures, are thought by many to be charismatic megafauna.

chelation. A method of removing certain heavy metals from the bloodstream, used especially in treating lead or mercury poisoning.

Christmas Bird Counts (CBCs). A winter monitoring program for birds in North America administered by the National Audubon Society. Each CBC consists of a "count circle," 24 km (15 mi) in diameter within which teams of observers count birds on a single day between 14 December and 5 January each year. The CBC began on Christmas Day 1900, when ornithologist Frank Chapman proposed the Christmas Bird Count as a new holiday tradition.

circle soaring. Soaring in circles. Birds often circle soar to remain within thermals.

cline. A gradation in one or more characteristics within a species, especially along a geographical range.

cold front. A large-scale, synoptic weather event in which cold, dense air passes through an area. In eastern North America, such movements are typically from northwest to southeast and are accompanied by northwesterly winds. In autumn, some of the best hawk and vulture flights occur following the passage of a cold front.

commensal relationship. A symbiotic relationship in which one species benefits from a common resource while the other species is not adversely affected. Vultures that benefit from humans are known as human commensals.

complete migrants. A species or population in which at least 90% of all individuals migrate regularly. A species or population, and not an individual, characteristic. *See also* irruptive migrants; partial migrants

contorted soaring. Vultures use this type of multiple-turning soaring mainly at low altitudes <150 ft (<45 m) when searching for carrion in small-scale, shear-induced turbulence within the atmospheric boundary layer.

convergent evolution. The independent evolution of similar behavioral, physiological, or anatomical traits in distantly related species. Diurnal raptors and owls share many traits derived by convergent evolution because of the similar niches they occupy.

corvids. A family of typically gregarious, small to large, black, black-and-white, or brightly colored birds that includes crows, ravens, jays, and magpies.

cosmopolitan. Occurring worldwide on all habitable land masses including at some islands. For raptors, habitable land masses include North and South America, Europe, Africa, Asia, Australia, and many islands, but not Antarctica. The Peregrine Falcon and the Osprey are considered cosmopolitan or near-cosmopolitan species. There are no cosmopolitan vultures.

Critically Endangered. A species facing an extremely high risk of extinction in the wild in the immediate future. *See also* Endangered; Vulnerable

cross winds. In birds, winds that intersect the preferred direction of travel at perpendicular or near perpendicular angles, and that often alter the direction of travel via a process called *wind drift*. *See also* head winds; tail winds

culmen. The upper ridge of a bird's bill.

DDT (dichlorodiphenyltrichloroethane). A synthetic organochlorine pesticide (OCP) once widely used in agriculture and human-disease control. A low-cost, persistent, broad-spectrum compound often termed a miracle pesticide, DDT came into widespread use in the late 1940s. However, its use in agriculture was banned in many countries, including Canada, United States, and much of western Europe, in the early 1970s when it was found to be a contaminant in human body tissue and in many ecosystems globally. DDT negatively affects the reproductive success of birds, including vultures, by disrupting a female's ability to produce sufficient eggshell material for her eggs.

DDT era. The time between the late 1940s and the early 1970s when DDT was in widespread use in North America and Western Europe and when populations of many birds in these areas, including Peregrine Falcons and Bald Eagles, declined precipitously.

deflection updrafts. Pockets of rising air created when horizontal winds encounter and deflect up and over mountains, ridges, escarpments, buildings, and even tall vegetation.

delayed maturation. Breeding for the first time, not in the first year of life, but in later years. May differ between sexes within species.

delayed return migration. Occurs when juvenile birds, including some vultures, remain on their wintering grounds during their entire second and, sometimes, third year before returning to the breeding grounds. Delayed return migration often happens in species with delayed maturation. The phenomenon is believed to save subadults the expense of migrating and to eliminate competition with adults during the breeding season.

dewlaps. A fold of loose skin hanging from the neck or throat.

diclofenac. A nonsteroidal, anti-inflammatory drug approved by the US Food and Drug Administration (US FDA) for use in humans and animals. Use in veterinary medicine, however, can kill vultures that consume carcasses of livestock treated with the drug. Even in small amounts, diclofenac is toxic to vultures, causing kidney failure.

dihedral. The upward angle of a bird's wings when soaring.

dispersal. Purposeful movement away from population centers that acts to separate members of populations. Often undertaken by recently fledged individuals, dispersal acts to increase population ranges, while reducing population densities overall.

distensible crop. An expandable, and when filled, externally visible sack on the throat side or ventral surface of the esophagus, capable of holding large amounts of recently consumed food. In many scavenging raptors, fully distended crops are thought to signal recent feeding by an individual.

diurnal. Active during daylight hours. Hawks, eagles, falcons, vultures, and their allies, are diurnal birds of prey. *Compare with* nocturnal

drag. A force that resists the motion of a body through a gas or liquid. Birds encounter three types of drag when flying: induced drag created by trailing vortexes, parasite drag created by the outline of a bird's body, and profile drag created by the bird's flapping wings. Induced drag decreases as speed increases, parasite drag starts at zero and increases as a cube of speed, and profile drag is constant across speeds.

dynamic (or sheer) soaring. Soaring that takes advantage of the vertical wind gradient that occurs close to a flat surface, when friction slows the layer of air in contact with the surface. Birds engaged in dynamic soaring alternately soar upward into increasing headwinds to gain altitude and then turn and glide downwind to gain airspeed, before again turning into the headwind and using the kinetic energy gained on the downwind leg to gain altitude on the upwind leg. Albatrosses and other pelagic seabirds are believed to use dynamic soaring while soaring at sea.

ecological equivalent. A species that can replace another species ecologically when the other species is absent or inactive. Owls are said to be ecological equivalents of diurnal birds of prey, and vice versa, because owls are active principally at night, whereas diurnal birds of prey are active principally during the day. In North America, Great Horned Owls are said to be ecological equivalents of Red-tailed Hawks.

ecological pyramid. (also known as food pyramid). A graphical representation of the food relationship of an ecological community, expressed quantitatively as numbers, mass, or energy at each tropic level, with producers at the bottom, and primary, secondary, and higher-level consumers ascending, in order, toward the top.

ectoparasites. Parasites such as fleas, flies, lice, mites, and ticks that live on the skin and feathers of birds.

Endangered. A species facing a very high risk of extinction in the near future. *See also* Critically Endangered; Vulnerable

endemics. Species that are native to, and are restricted to, a relatively small geographic region. *Island endemics* are species limited to a single island or archipelago. Endemics are often called *limited-range species.*

engorged crop. A crop that is filled to excess. In birds, the crop is a dilation of the lower esophagus that stores food.

essential elements. Biologically necessary elements used in small amounts by a variety of organisms.

exploitative competition. Occurs when individuals compete for access to a common limiting resource when the use of that resource by one individual depletes the amount of it available to other individuals.

extra-pair copulations. Copulations with a bird other than one's mate.

extinction. Total elimination of a species throughout its entire range.

extirpation. Total elimination of a species from a part, but not all of its range. Not to be confused with *extinction.*

eye ring. Distinctively colored skin or feathering surrounding the eye.

facultative scavenger. A species that feeds on both living prey and carrion, depending on the relative availability of the two resources, but does not depend on either food resource entirely.

falcons. Members of the genus *Falco*, a group of 37 diurnal raptors with long, pointed wings and long tails.

fat soluble. The ability to dissolve in fat. Pesticides that are fat soluble, but water insoluble, accumulate in the fatty tissues of animals and can result in *biological magnification* of pesticides in other species in the food chain, including both predatory and scavenging raptors.

fledge. The act of a nestling when it first flies from the nest.

flex gliding. High-speed gliding during which a bird reduces its wingspan, wing slotting, and overall flight surface by flexing its wings inward and toward its body. Flex gliding increases aerodynamic performance during high-speed gliding flight.

flight feathers. The long, stiff feathers of the wings and tail. On the wings they are called *remiges* and include both the primaries and secondaries. On the tail they are called *rectrices*.

flocking. Joining together in groups as a result of social attractions. Flocks differ from aggregations in that the latter result when birds are attracted to locations by physical or ecological factors alone.

food chain. A sequence of organisms including primary producers (mainly plants), herbivores (plant eaters), and carnivores (meat eaters), through which energy moves in ecosystems.

gang behavior. Occurs when groups of conspecific or congeneric individuals feed ravenously together at carcasses in feeding frenzies that are typically avoided by other species of scavengers, even when the latter typically dominate the former during one-on-one encounters.

gape. The opening formed when the mouth is open wide.

generation time. The average age of breeding females within a population.

genus. A group of one or more closely related species.

glide ratio. The ratio of sinking speed to forward speed in a glide.

grit. Small, loose particles of stone or sand consumed by birds to assist grinding food in the gizzard.

gular fluttering. Heat-dissipation mechanism in which an individual bird vibrates its wetted gular (or throat) surface with its mouth open. Particularly common in nestlings.

habitat. A species habitat is the ecological location where a species can be found, or its "address" in nature. *See also* niche

hacking. A training method in which humans help young birds of prey by providing them with food, exercise, and experience. This technique is used in vultures to prepare them to recognize and feed successfully at carcasses in groups. In vultures, hacking includes raising individuals in captivity and releasing them in the wild at carcasses where other vultures are feeding.

half-life. The time it takes for half the mass of a substance, including a synthetic pesticide, to degrade in the environment into its by-products and, therefore, lose initial effect.

head winds. Winds aligned against the preferred direction of travel that strike birds in the head and hinder them by pushing against them and slowing down their ground speed, or rate of travel. *See also* cross winds; tailwinds

high aspect-ratio wing. High aspect-ratio wings are long, narrow, and pointed. They occur principally in birds that live in open habitats and that often fly at high speeds.

histocompatibility complex. A large locus on vertebrate DNA containing a set of closely linked polymorphic genes essential for an adaptive immune system.

Holocene. The latest interval of geologic time, covering the past 11,700 years of Earth's history.

homeotherm. Warm-blooded. An animal that is able to keep its internal body temperature constant as ambient temperatures vary. Raptors are homeotherms. *See also* poikilotherm

human commensal. A species that benefits from the presence of humans without negatively affecting them.

human subsidies. An ecological condition in which human activities (e.g., herding, farming, disposal of organic trash, etc.) provide food for otherwise nutritionally stressed wildlife populations, including those of birds of prey.

Humboldt Current. The Humboldt Current, also called the Peru Current, is a cold, low-salinity ocean current that flows north along the western coast of South America. It is an eastern boundary current flowing in the direction of the equator and extends 310–620 mi (500–1000 km) offshore. Upwelling associated with the current involves wind-driven motion of dense, cooler, and usually nutrient-rich water toward the ocean surface, replacing the warmer, usually nutrient-depleted surface water. The nutrient-rich upwelled water stimulates the growth and reproduction of primary producers such as phytoplankton that, in turn, support high populations of consumers including anchoveta that serve as a food resource for seabirds. Because of the biomass of phytoplankton and presence of cool water in these regions, upwelling zones can be identified by cool sea surface temperatures and high concentrations of chlorophyll-a.

information center. The information center hypothesis suggests that communal roosts and colonial breeding sites function as information centers for resource availability, with unsuccessful hunters at such sites watching for successful hunters to return and then departing in the direction they have returned from.

insular. An island form. A species that is found on islands, but not on the mainland.

interference competition. Interference competition occurs directly between individuals via interference with foraging, survival, reproduction of others, or by preventing an individual's presence in a part of a habitat.

irruptive migrants. Species or regional populations in which the extent of migratory movement varies annually, typically caused by among-year shifts in prey abundance, and whose migrations are less regular than those of partial and complete migrants. A species or population, not an individual, characteristic. *See also* complete migrants and partial migrants

island endemic. *See* endemics

KDE. The acronym for a **k**ernel **d**ensity **e**stimation, in which KDE is a nonparametric estimate of the probability density function of a bird's occurrence in a location. KDE provides a more conservative (smaller) estimate of an individual's home range than an MCP (a minimum convex polygon). *See also* MCP

kettle. Among bird-watchers, a flock of birds, especially when individuals are circling in a thermal. Vultures in kettles are sometimes said to be *kettling*. Both terms owe their origins to a boulder field called The Kettle at Hawk Mountain Sanctuary in eastern Pennsylvania, USA, over which migrating Broad-winged Hawks and other raptors often circle upward while soaring thermals that are created by the dry, un-vegetated surface of the field.

key innovation. An evolutionary change in an organism that significantly increases access to ecological resources that subsequently enhance range expansion, population growth, and, in some instances, adaptive radiation. Lungs, which developed in aquatic vertebrates, enabled the rapid diversification of land-based vertebrate life. Feathers, which developed in reptiles, enabled the rapid radiation of flighted birds. Both are considered key innovations. The dihedral wing configuration in vultures enhances within-boundary-layer soaring, and is believed to have enhanced range expansion, population growth, and adaptive radiation.

kernel density estimate (KDE). Bivariate, Gaussian, and normal distribution kernel density estimates (KDEs) are widely used for determining the sizes of animal home ranges based on location data. *See also* KDE

keystone species. Species that have a major influence on the community structure of an ecosystem.

Kleiber's Law. Named after Max Kleiber for his biological work in the early 1930s, Kleiber's Law is the observation that for most animals, including vultures, metabolic rate scales to the ¾ power of the individual's body mass.

kleptoparasite. A bird or other animal that habitually robs animals of other species of their food.

leap-frog migration. Occurs when migratory populations that breed at high latitudes migrate substantially farther and "leap-over" migratory and nonmigratory populations breeding at lower latitudes, thereby reversing their latitudinal relationship between seasons. *See also* chain migration

linear soaring. Long-distance, straight-line soaring in *thermal streets*, a phenomenon more typical of tropical than temperate regions.

lores. The area between the eye and upper beak. Featherless in some vultures.

melanin. A class of polymers derived from the amino acid tyrosine that functions as blackish pigments in birds and other animals.

melanistic. An individual showing high levels of melanin in its feathering, resulting in a dark color morph.

mesocarp. The middle layer of the pericarp of a fruit, such as the flesh of a peach.

metabolic water. Water created inside a living organism through their metabolism via oxidation of their food. Animal metabolism produces about 110 g of water per 100 g of fat, 42 g of water per 100 g of protein, and 60 g of water per 100 g of carbohydrate.

metabolism. All the synthetic (anabolic) and degradative (catabolic) chemical processes of living organisms necessary to sustain life.

MCP. The acronym for a **m**inimum **c**onvex **p**olygon, in which the MCP is the smallest polygon around location points with all interior angles less than 180 degrees. MCPs are common estimators of home ranges, but they can potentially include areas not used by the animal, and as such, overestimate the home range. *See also* KDE; minimum convex polygon

migration bottleneck. A site where migration corridors converge and subsequently diverge, and at which large numbers of migrants concentrate. Mountain passes, isthmuses, narrow coastal plains, and water crossings, including the Isthmus of Panama, the Strait of Gibraltar, and the Bosporus, are examples of migration bottlenecks. Major migration bottlenecks create funnel-shaped migration flyways.

migration connectivity. The ecological phenomenon in which migratory individuals that breed close to one another on the breeding grounds also overwinter close together on their wintering grounds.

migration corridors (or flyways). Routes used by massive numbers of migrating raptors that are created when lines and diversion lines act to concentrate migrants.

mimicry. The close resemblance of one species (the mimic) to a second species (the model) to deceive one or more other species (the operators).

minimum power speed. The speed at which a bird, including vultures and condors, consumes the least amount of energy per unit time; that is, the speed that allows it to say airborne longest.

minimum convex polygon (MCP). The MCP is one of the simplest estimates of an animal's home range from location data. Although in widespread use, its limitations include often overestimating the size of a home range. *See also* KDE

molecular phylogenetics. Analyses of hereditary molecular differences, principally of DNA sequences, to learn about an organism's ancestral relationships.

monogamy. A social system in which a single male and a single female form a breeding relationship. The individuals involved are said to be monogamous.

monophyletic group. A group of species descended from a single ancestral species. Usually used in the restricted sense to mean the ancestral species and all of its descendants.

mountain (or slope) updrafts. Pockets of rising air created when horizontal winds encounter and deflect up and over mountains, ridges, escarpments, buildings, and even tall vegetation.

narrow-front migration. Migration in which dispersed migrants deviate from their initial directions, either because they are attracted to or are avoiding certain geographic features such as mountain ranges, river systems, and coastlines, or because they seek out and join conspecifics en route. Narrow-front migration results in concentrated movements of migrants.

navigation. Movement toward a goal that often requires reorientation en route and includes knowledge of the distance between the present location and the goal.

NDVI. Acronym for **n**ormalized **d**ifference **v**egetation **i**ndex. Based on satellite infrared imagery, NDVI measures primary productivity (i.e., photosynthetic activity of a geographic area).

Neogene. A geologic period and system 20–23 million years ago.

New World. North America, Central America, South America, and associated islands including Greenland. Also known as the Western Hemisphere. *See also* Old World

niche. The ecological role of a species in a community, or the so-called ecological profession of the species. Some species' niches are more specialized than others. *See also* habitat

nocturnal. Active during the hours of darkness. Owls are nocturnal birds of prey. *See also* diurnal

nomadic. Wandering, sometimes seasonally, over large areas, without specific directional components.

normalized difference vegetation index (NDVI). The NDVI is a graphic indicator used to analyze remote sensing measurements, often from satellites in space, for assessing the extent of living (green) vegetation in a mapped area.

nursery area. An area of low adult population density where recently fledged, juvenile young and subadult vultures develop adult food-searching, courting, and breeding skills needed for successful adulthood.

obligate scavenging bird of prey. A raptor that depends on carrion to survive. Vultures are obligate scavenging birds of prey.

obligate soaring migrant. A migratory raptor, including vultures, that depends on the use of energy-efficient soaring flight to complete its migratory journeys.

oceanic islands. Islands formed by volcanic activity from the floor of the ocean that have never been attached to a continent.

Old World. Europe, Asia, Africa, Australia, and associated islands. Also known as the Eastern Hemisphere. *See also* New World

organochlorine pesticides. Also known as chlorinated hydrocarbons, organochlorines are a class of synthetic, broad-spectrum, contact pesticides that include DDT and dieldrin, which first came into widespread use in the 1940s and 1950s. Usually used against insect pests, organochlorines resist biodegradation, persist for long periods in ecosystems, and often accumulate in non-target organisms including vultures.

orographic. Relating to mountains, especially with regard to their position and form.

outbound migration. Migratory movements occurring immediately after the breeding season. *See also* return migration

paleospecies. A species known only from fossils.

panmictic. Random mating within a breeding population.

partial migrants. A species or a population in which fewer than 90% of all individuals regularly migrate. A species or population, and not an individual, characteristic. *See also* complete migrants; irruptive migrants

patagium (patagial). A fold of vascularized skin on the leading edge of the inner wing of a bird connecting the shoulder to the wrist.

peristaltic egestion. Alternating radially symmetrical contraction and relaxation of esophageal muscles that propagate a series of waves from the stomach to the throat, thereby permitting raptors to regurgitate boluses of indigestible waste from their stomachs.

persistent. Among pesticides, those having a relatively long half-life. Synthetic organochlorine pesticides, in particular, are persistent pesticides. DDT, for example, has a half-life of 15 years.

philopatric. Tending to return to or remain near one's birthplace.

photosynthesis. The biochemical process that uses the radiant energy in sunlight to synthesize carbohydrates from carbon dioxide and water in the presence of chlorophyll. Green plants engage in photosynthesis.

phylogenetic extinction. The termination of a phyletic lineage (i.e., species) without the formation any descendent lineage. Sometimes called *terminal extinction*.

phylogenetic inertia. Limits on future evolutionary pathways imposed by previous adaptations.

phylogeny. The evolutionary history of a group. A species', or group of species', ancestral relationships.

Pleistocene. The geological epoch from about 2,600,000 to 11,700 years ago, spanning the world's most recent repeated glaciations.

poikilotherm. Cold-blooded. An animal that is largely or totally unable to keep its internal body temperature constant as ambient temperatures vary. *See also* homeotherm

polyandrous. A polygamous situation known as polyandry, in which one female simultaneously has two or more male mates.

polygamous. A mating situation known as polygamy, in which one individual simultaneously has two or more mates.

polygynous. A polygamous situation known as polygyny, in which one male simultaneously has two or more female mates.

polyphyletic. A group that is derived from two or more distinct lineages, rather than from a common ancestor.

precocial. Young birds that hatch relatively well developed, with eyes open, thick down, and the capacity to walk.

primaries. The outer-most flight feathers on the wing. All raptors, including vultures, have 10 of these.

propagules. The minimum number of individuals of a species needed for colonization.

proventriculus. The upper part of the bird's two-part stomach that secretes mucus, hydrochloric acid, and enzymes to help chemically digest food.

regurgitation pellets. As is true of many birds of prey, vultures routinely regurgitate mucous-covered pellets of the less digestible remains of their food. Pellets typically are regurgitated around dawn or near evening.

return migration. Migratory movements occurring immediately before the breeding season. *See also* outbound migration

reciprocal migration. Migration in which the migration of a behaviorally dominant species into an area results in the migration of a subordinate species.

reverse migration. Migratory movement in the direction opposite of that typical of the season.

reverse size dimorphism (RSD). When the females of a species are larger than their male counterparts. RSD occurs in many predatory birds including skuas, frigatebirds, boobies, and most raptors (including owls). The function of RSD remains unsettled.

saprophytic. A plant, fungus, or microorganism that lives on dead or decaying organic matter.

satellite tracking. Following the movements of animals, including raptors, to which a UHF (ultra-high-frequency) transponder has been attached (satellite tagging). Used together with an array of satellites, these tracking devices can be used to monitor the movements of vultures across the surface of the globe.

scramble competition. Direct competition for a limiting resource, in which competitors scramble for access to the resource with some, but not all, competitors gaining access to the resource.

secondaries. The inner-most flight feathers on a bird's wing. In raptors, including vultures, the number of secondaries varies among species.

shear soaring. Soaring related to wind shear, also called *dynamic soaring*, is a technique used by soaring birds to maintain flight without wing flapping. If the wind shear (i.e., vertical or spatial variation in wind velocity) is of sufficient magnitude, a bird can climb into the wind gradient, trading ground speed for height, while maintaining airspeed. By then turning downwind, and diving at increasing speed through the wind gradient, they gain potential energy.

shell midden. An old human dump, usually for bivalve shells.

short-stopping. A phenomenon first described in migratory waterfowl that occurs when migrants shorten the lengths of their outbound movements to take advantage of newly available wintering areas that are closer to their breeding grounds than traditional sites.

siblicide. The killing of an organism by one of its siblings. In vultures, siblicide occurs when a nestling kills one of its nest mates.

slope-soaring. Soaring in updrafts that are created when horizontal winds strike and are deflected over isolated mountains and mountain ranges.

slope updrafts. Updrafts produced when wind is deflected upward when it strikes a slope.

slotted wingtips. Also called *emarginated wingtips*, these are wingtips in which the outermost primaries or flight feathers, when fully spread, create gaps in the margin of the wingtip, giving the tip a fingered appearance. Species with slotted wingtips have deeply notched outer-most primaries that are wide at the base but that narrow abruptly about halfway to the tip and, therefore, are much narrower at the tip. Although their function remains unsettled, many scientists believe slotting reduces drag near the wingtips by creating a series of mini-airfoils.

soaring. Level or ascending non-flapping flight on outstretched wings in which upward air movements are equal to or greater than the bird's rate of descent in a glide.

species. A group of interbreeding organisms that is reproductively isolated from other groups of organisms.

subnivean. Occurring under the snow.

sunning. Exposing plumage to the sun, usually by spreading the wings and tail, erecting back feathers, and remaining motionless.

superciliary. Of or pertaining to the eyebrow.

super-flocking migrants. Long-distance, typically trans-equatorial migrants that travel in large flocks of hundreds to tens of thousands of birds, particularly in the tropics and subtropics. Some populations of Turkey Vultures are super-flocking migrants.

surface albedo. A percentage measure of surface reflectivity.

symbiotic relationship. A close and long-term biological interaction between two taxa, including *mutualism* in which both species benefit, *commensalism* in which one but not both species benefit, or *parasitism* in which one species benefits and the other is harmed.

syringeal. Having to do with the avian syrinx, the sound-producing organ of birds, or song box.

tailwinds. Winds aligned with the preferred direction of travel. Tailwinds help raptors by increasing their ground speed, or rate of travel, by "pushing" them forward in the preferred direction of travel. *See also* cross winds; head winds

talon grappling. Talon grappling is a mid-flight physical interaction between vultures and other raptors characterized by two birds locked together by their talons and spinning as they fall.

talons. The deeply curved, grasping claws of a raptor, including some vultures.

taxonomic principle of priority. The principle of recognizing the first valid application of a name to an organism.

temperate zone. Latitudinally, the temperate regions between the tropics of Cancer and Capricorn (23.5° north and south of the equator) and the polar zones (greater than 66.5° north and south). Climates in the temperate zone are characterized as moderate and not usually excessive or extreme.

Teratorns. A group of very large, extinct vulturine birds that lived in the Americas until the Pleistocene. Members of the avian subfamily Teratornithinae.

termite alates. A termite in its winged form.

thermal. A pocket of warm, rising air created by the differential heating of the earth's surface.

thermal soaring. Soaring flight in thermals, or pockets of warm rising air. Vultures often circle when soaring in thermals to remain within them.

thermal streets. Linear arrays of thermals that are often aligned with prevailing winds. When thermal streets are oriented along the so-called principal axis of migration they enable migrants to soar linearly within them in much the same way as migrants slope soar.

thermoregulation. The ability of an animal to keep its body temperature within certain limits. Most raptors, including vultures and condors, thermoregulate their internal body temperature at between approximately 39.5°C in large species and 41°C in small species.

tool use. The use of physical objects other than the animal's own body or appendages as a means to extend the physical influence realized by the animal. Egyptian Vultures are known to use stones as part of their feeding behavior.

torpor. Metabolic state in which the core body temperature of a homeotherm drops several degrees and metabolic processes slow.

trade winds. Predominant northeasterly and southeasterly subtropical and tropical oceanic winds that encircle the globe 5° to 30° north and south of the equator, respectively. Thermal convection produced by these winds creates sea thermals.

trade-wind thermals. Sea thermals created when predominant northeasterly winds in the northern hemisphere trade-wind zone (i.e., 5° to 30° N) and predominant southeasterly winds in the southern hemisphere trade-wind zone pass over increasingly warmer subtropical and tropical waters. Raptors sometimes take advantage of such over-water thermals when migrating in the trade-wind zones north and south of the equator.

trade-wind zone. *See* trade-wind thermals.

transhumance; transhumant. The seasonal movement of livestock between mountain and lowland pastures under the care of herders.

trap shy. A trap-shy animal is one that having been trapped recognizes and shies away from traps.

trophic guild. A group of species that exploits the same resources, or that exploits different resources in related ways. Species within a guild do not necessarily occupy the same, or even similar, ecological niches.

tropics. The geographical zone between the Tropic of Cancer (23.5° North) and the Tropic of Capricorn (23.5° South). The tropics include all of the areas on Earth where the Sun is directly overhead at least once a year. Tropics tend to have hot and humid climates.

urohidrosis. Defecating uric acid on one's legs.

vagrant. In birds, a species that does not regularly breed, overwinter, or migrate through an area, but occurs there infrequently, presumably when lost.

velocity of minimum power. The flapping-flight speed at which a raptor expends the minimum amount of energy per unit of time. This speed allows a raptor to remain in the air the longest in flapping flight before running out of metabolic energy.

visual acuity. The ability of an eye to resolve details.

voles. Mouse-like small rodents with stout bodies, smallish eyes and ears, and a short tail. Voles are frequently preyed upon by raptors in the northern temperate zone.

Vulnerable. A species facing a high risk of extinction in the wild in the medium-term future. *See also* Critically Endangered; Endangered

vulture restaurants. Supplemental feeding sites for vultures at which livestock carcasses, offal, and so forth are made available to vultures on a regular basis.

Wallacea. Indonesian islands separated by deep-water straits from the Asian and Australian continental shelves, including Sulawesi, the largest island in the group, Lombok, Sumbawa, Flores, Sumba, Timor, Halmahera, Buru, Seram, and many smaller islands.

wing dihedral. The upward angle of a bird's wings when soaring.

wind drift. When migrants encountering crosswinds are pushed off their intended course even while maintaining the same heading.

wingspan. The distance from wingtip to wingtip on fully extended wings.

wingtip. The distal end of a bird's wing.

wing chord. The distance between the carpal joint and the wingtip of a bird's wing.

wing loading. A bird's mass divided by the combined area of its outstretched wings. Species with low wing loading have relatively large wings for their body mass. Those with high wing loading have relatively small wings. Small birds tend to have lighter wing loading than larger birds. For their body mass, raptors tend to be more lightly wing loaded than other birds.

wing slotting. Occurs when the distal ends of the outermost primaries on a bird's wing are separated horizontally and vertically in flight, thereby creating separate aerodynamic surfaces. Wing slotting is enhanced in some species by the presence of emarginate, or notched, outermost primaries. Wing theory suggests that wing slotting increases the aerodynamic wingspan of a bird, and reduces wingtip drag at low speeds.

REFERENCES AND RECOMMENDED READINGS BY CHAPTER

Introduction: Origins and Evolution of Vultures

Arad, Z., and M. H. Bernstein. 1988. Temperature regulation in Turkey Vultures. Condor 90:913–919.

Bahat, O., and I. Choshniak. 1998. Nocturnal variation in body temperature of Griffon Vultures. Condor 100:168–171.

Campbell, K. E., Jr., and E. P. Tonni. 1983. Size and locomotion in teratorns (Aves: Teratornithidae). Auk 100:390–403.

Cracraft, J., and P. V. Rich. 1972. The systematics and evolution of the Cathartidae in the Old World Tertiary. Condor 74:272–283.

Feduccia, A., 1974. Another Old World vulture from the New World. Wilson Bull. 86:251–255.

Feduccia, A. 1996. The origin and evolution of birds. New Haven: Yale University Press.

Gadow, H. 1892. On the classification of birds. Proc. Zool. Soc. Lond. 1892:229–256.

Holtz, T. R., Jr. 2008. A critical reappraisal of the obligate scavenging hypothesis for *Tyrannosaurus rex* and other tyrant dinosaurs. In *Tyrannosaurus rex*, ed. P. Larson and K. Carpenter. Bloomington: Indiana University Press.

Horner, J. R. 1994. Steak knives, beady eyes, and tiny little arms (a portrait of *Tyrannosaurus rex* as a scavenger). *In* Dino Fest Proceedings, ed. G. Rosenberg and D. Wolberg. Paleontological Soc. Special Publ. 7.

Howard, H. 1972. The incredible teratorn again. Condor 74:341–344.

Hertel, F. 1994. Diversity in body size and feeding morphology within past and present vulture assemblages. Ecology 75:1074–1084.

Jackson, A. L., G. D. Ruxton, and D. C. Houston. 2008. The effect of social facilitation on foraging success in vulture: a modelling study. Biol. Letters 4:311–313.

Johnson, J. A., H. R. L. Lerner, P. C. Rasmussen, and D. P. Mindell. 2006. Systematics within *Gyps* vultures: a clade at risk. BMC Evol. Biol. 6:65. doi.org/10.1186/1471-2148-6-65.

Lerner, H. R. L., and D. P. Mindell. 2005. Phylogeny of eagles, Old World vultures, and other Accipitridae based on nuclear and mitochondrial DNA. Mol. Phylogenet. Evol. 37:327–346.

Mayr, E. 1946. History of North American bird fauna. Wilson Bull. 58:1–68.

Miller, A. H., and L. V. Compton. 1939. Two fossil birds from the lower Miocene of South Dakota. Condor 61:153–156.
Miller, L. H. 1909. Teratornis: a new avian genus. University California Bull. Dept. Geol. 5:305–317.
Miller, L. H. 1910. The condor-like vultures of Rancho La Brea. University California Bull. Dept. Geol. 6:1–19.
Miller L. H., and I. S. DeMay. 1942. The fossil birds of California; and avifauna and bibliography with annotations. University California Publ. Zool. 47:47–142.
Mindell, D. P., J. Fuchs, and J. A. Johnson. 2018. Phylogeny, taxonomy, and geographic diversity of diurnal raptors: Falconiformes, Accipitriformes, and Cathartiformes. *In* Birds of prey, ed. J. H. Sarasola, J. M. Grande, and J. J. Negro. Cham, Switzerland: Springer Nature.
Olsen, S. L., and H. F. Alvarenga. 2002. A new genus of small teratorn from the middle Tertiary of the Taubaté Basin, Brazil (Aves: Teratornithidae). Proc. Biol. Soc. Washington 115:701–705.
Rich, P. V. 1980. Preliminary report on the fossil avian remains from late Tertiary sediments at Langebaanweg (Cape Province), South Africa. S. A. J. Science 76:166–170.
Rich, P. V. 1983. The fossil history of vultures: a world perspective. *In* Vulture biology and management, ed. S. R. Wilbur and J. A. Jackson. Berkeley: University of California Press.
Ruxton, G. D., and D. C. Houston. 2002. Could *Tyrannosaurus rex* have been a scavenger rather than a predator? An energetics approach. Proc. R. Soc. Lond. B 270:731–733.
Ruxton, G. D., and D. C. Houston. 2004. Obligate vertebrate scavengers must be large soaring fliers. J. Theoretical Biol. 228:431–436.
Schmidt-Nielson, K. 1972. Energy cost of swimming, running, and flying. Science 177:222–228.
Seibold, I., and A. J. Helbig. 1995. Evolutionary history of Old and New World vultures inferred from nucleotide sequences of the mitochondrial cytochrome *b* gene. Philos. Trans. R. Soc. Lond. B 350:163–178.
Thorstrom, R., and L-A. R. de Roland. 2000. First nest-description, breeding behavior and distribution of the Madagascar Serpent-Eagle *Eutriorchis astur*. Ibis 142:217–224.
Zhang, Z., Y. Huang, H. F. James, and L. Hou. 2012. Two Old World vultures from the middle Pleistocene of northeastern China and their implications for interspecific competition and biogeography of Aegypiinae. J. Vert. Palaeontol. 32:117–124.
Zhiheng, L., J. A. Clarke, Z. Zhou, and T. Deng. 2016. A new Old World vulture from the late Miocene of China sheds light on Neogene shifts in the past diversity and distribution of the Gypaetinae. Auk 133:615–625.

Chapter 1: Essential Ecology of Scavengers

Allee, W. C., A. E. Emerson, O. Park, T. Park, and K. P. Smith. 1949. Principles of animal ecology. Philadelphia: Saunders.
Bildstein, K. L. 2017. Raptors: the curious natural history of diurnal birds of prey. Ithaca: Cornell University Press.
DeVault, T. L., O. E. Rhodes, and J. A. Shivik. 2003. Scavenging by vertebrates: behavioral, ecological, and evolutionary perspectives on an important energy transfer pathway in terrestrial ecosystems. Oikos 102:225–234.
Elton, C. 1947. Animal ecology. New York: Macmillan.
Janzen, D. H. 1977. Why fruits rot, seeds mold, and meat spoils. American Naturalist 111:691–713.

Magoun, A. J. 1976. Summer scavenging activity in northeastern Alaska. M.S. Thesis. Fairbanks: University of Alaska.
Mullen, D. A., and F. A. Pitelka. 1972. Efficiency of winter scavengers in the Arctic. Arctic 25:225–231.
Odum, E. 1959. Fundamentals of ecology. Second edition. Philadelphia: Saunders.
Prince, P. A. and R. A. Morgan. 1987. Diet and feeding ecology of Procellariformes. *In* Seabirds: feeding biology and role in marine ecosystems, ed. J. P. Croxall. Cambridge: Cambridge University Press.
Putman, R. J. 1983. Carrion and dung: the decomposition of animal wastes. London: Edward Arnold.

Chapter 2: Species Descriptions and Life Histories

General References

BirdLife International. 2019. https://www.birdlife.org. Cambridge, United Kingdom.
Gill, F., and M. Wright. 2006. Birds of the world: recommended English names. Princeton: Princeton University Press.
Ruxton, G. D., and D. C. Houston. 2004. Obligate vertebrate scavengers must be large soaring fliers. J. Theoretical Biol. 228:431–436.

Black Vulture

Bildstein, K. L., M. J. Bechard, P. Porras, E. Campo, and C. J. Farmer. 2007. Seasonal abundances and distributions of Black Vultures (*Coragyps atratus*) and Turkey Vultures (*Cathartes aura*) in Costa Rica and Panama: evidence for reciprocal migration in the Neotropics. *In* Neotropical Raptors, ed. K. L. Bildstein, D. R. Barber, and A. Zimmerman. Kempton: Hawk Mountain Sanctuary.
Buckley, N. J. 1999. Black Vulture (*Coragyps atratus*). *In* Birds of North America, No. 411, ed. A. Poole and F. Gill. Philadelphia: Birds of North America.
Carrete, A., S. A. Lambertucci, K. Speziale, O. Ceballos, A. Taviani, M. Delibes, F. Hiraldo, and J. A. Donazar. 2010. Winners and losers in human-made habitats: interspecific competition outcomes in two Neotropical vultures. Anim. Conserv. 13:390–398.
Darwin, C. 1839. The Voyage of the Beagle. New York: Random House.
Eisenmann, E. 1963. Is the Black Vulture migratory? Wilson Bull. 75:244–249.
Henckel, E. 1976. Turkey Vulture banding problem. N. Am Bird Band. 1:126.
Larochelle, J., J. Delson, and K. Schmidt-Nielsen. 1982. Temperature regulation in the Black Vulture. Canadian J. Zool. 60:491–494.
Parker, P. G., T. A. Waite, and M. D. Decker. 1995. Kinship and association in communally roosting Black Vultures. Anim. Behav. 49:395–401.
Parmalee, P. W. 1954. The vultures: their movements, economic status, and control in Texas. Auk 71:443–453.
Parmalee, P. W., and B. G. Parmalee. 1967. Results of banding studies of the Black Vulture in eastern North America. Condor 69:146–155.
Pennycuick, C. J. 1983. Thermal soaring compared in three dissimilar tropical bird species, *Fregata magnificens*, *Pelecanus occidentalis*, and *Coragyps atratus*. J. Exp. Biol. 102:307–325.

Rabenold, P. P. 1986. Family associations in communally roosting Black Vultures. Auk 103:32–41.
Rabenold, P. P. 1987. Recruitment to food in Black Vultures: evidence for following from communal roosts. Anim. Behav. 35:1775–1785.
Rabenold, P. P. 1987. Roost attendance and aggression in Black Vultures. Auk 104: 647–653.
Rabenold, P., and M. D. Decker. 1990. Black Vultures in North Carolina: statewide population studies and analysis of Chatham County population trends. Raleigh: N. Carolina Wildl. Resour. Comm.
Skutch, A. F. 1969. Notes on the possible migration and the nesting of the Black Vulture in Central America. Auk 86:726–731.
Wilson, A. 1840. Wilson's American Ornithology. Boston: Otis, Broaders.

Turkey Vulture

Bildstein, K. L. 2006. Migrating raptors of the world. Ithaca: Cornell University Press.
Coleman, J. S., and J. D. Fraser. 1987. Food habits of Black and Turkey Vultures in Pennsylvania and Maryland. J. Wildl. Manage. 51:733–739.
DeVault, T. L., B. D. Reinhart, I. L. Brisbin, and O. E. Rhodes, Jr. 2004. Condor 106:706–711.
Graña Grilli, M., K. L. Bildstein, and S. A. Lambertucci. 2019. Nature's clean-up crew: quantifying ecosystem services offered by a migratory avian scavenger on a continental scale. Ecosystem Serv. doi:100990.
Hatch, D. E. 1970. Energy conserving and heat dissipating mechanisms of the Turkey Vulture. Auk 87:111–214.
Hedlin, E. M., C. S. Houston, P. D. McLoughlin, M. J. Bechard, M. J. Stoffel, D. R. Barber, and K. L. Bildstein. 2013. Winter ranges of migratory Turkey Vultures in Venezuela. Wilson J. Ornithol. 47:145–152.
Holland, A. E., M. E. Byrne, A. L. Bryan, R. DeVault, O. E. Rhodes, Jr., and J. C. Beasley. 2017. Fine-scale assessment of home ranges and activity patterns for resident black vultures (*Coragyps atratus*) and turkey vultures (*Cathartes aura*). PLoS ONE 10.1371.
Houston, C. S., and P. H. Bloom. 2005. Turkey Vulture marking history: the switch from leg bands to patagial tags. N. Am. Bird Band. 30:59–64.
Houston, C. S, P. D. McLoughlin, J. T. Mandel, M. J. Bechard, M. J. Stoffel, D. R. Barber, and K. L. Bildstein. 2011. Breeding home ranges of migratory Turkey Vultures near their northern limit. Wilson J. Ornithol. 123:472–478.
Houston, C. S., B. Terry, M. Bloom, and M. J. Stoffel. 2007. Turkey Vulture nest success in abandoned houses in Saskatchewan. Wilson J. Ornithol. 119:742–747.
Kelly, N. E., D. W. Sparks, T. L. DeVault, and O. E. Rhodes, Jr. 2007. Diet of Black and Turkey Vultures in a forested landscape. Wilson J. Ornithol 119:267–270.
Kirk D. A., and M. J. Mossman. 1998. Turkey Vulture (*Cathartes aura*). *In* The birds of North America, No. 339, ed. A. Poole and F. Gill. Philadelphia: Birds of North America.
Rexer-Huber, K., and K. L. Bildstein. 2013. Winter diet of striated caracara *Phalcoboenus australis* (Aves, Polyborinae) at a farm settlement on the Falkland Islands. Polar Biol. 36:437–443.
Stewart, P. A. 1977. Migratory movements and mortality rate of Turkey Vultures. Bird-Banding 48:122–124.
Yahner, R. H., G. L. Storm, and A. L. Wright. 1986. Winter diets of vultures in south-central Pennsylvania. Wilson Bull. 98:157–160.

Lesser Yellow-headed Vulture

Belton, W. 1984. Birds of Rio Grande do Sol, Brazil, part 1. Rheidae through Furnariidae. Bull. Am. Mus. Nat. Hist. 178:1-437.

Campbell, M. O'N. 2015. Vultures: their evoluation, ecology, and conservation. Boca Raton: CRC Press.

Di Giacomo, A. G., and S. F. Krapovickas, eds. 2005. Historia natural y paisaje de la Reserva El Bagual. Temas de Naturaleza y Conservación No. 4. Buenos Aires: Aves Argentinas/Asociación Ornithologic del Plata.

Eitniear, J. C. 1985. Notes on the relative abundance and distribution of the lesser Yellow-headed Vulture in Mexico and Belize, Central America. Bull. World Work. Grp. Birds of Prey 2:17–20.

Eitniear, J. C., and S. M. McGehee. 2017. Lesser Yellow-headed Vulture mandibular ecomorphology and feeding interactions at an established feeding site in Belize. Texas J. Sci. 69:39–48.

Graves, G. R. 2016. Head color and caruncles of sympatric *Cathartes* vultures (Aves: Cathartidae) in Guyana and their possible function in intra- and interspecific signaling. Proc. Biol. Soc. Washington 129:66–75.

Houston, D. C. 1994. Observation on Greater Yellow-headed Vulture *Cathartes melambrotus* and other *Cathartes* species as scvengers in forest in Venezuela. Raptor conservation today, ed. B.-U. Meyburg and R. D. Chancellor. London: World Work. Grp. Birds of Prey and Owls.

Koester, F. 1982. Observations on migratory Turkey Vultures and Lesser Yellow-headed Vultures in northern Colombia. Auk 99:372–375.

Meijer, H. J. M., T. Sutikna, E. W. Sptomo, R. D. Awe, Jatmiko, S. Wasisto, H. F. James, M. J. Morwood, and M. W. Tocheri. 2013. Late Pleistocene-Holocene non-Passerine avifauna of Liang Bua (Flores, Indonesia). J. Vert. Paleontol. 33:877–894.

Rodríguez-Estrella, R. 1991. Lesser Yellow-headed Vulture (*Cathartes burrovianus*) capturing live prey in Veveracuz, México. Brenesia 34:159–160.

Severo-Neto, F. S. Paulino de Faria, and D. J. Santana. 2014. Adding some poison to menu: first report of a cathartid vulture preying on a venemous snake. Herpetol. Notes 7:675–677.

Wetmore, A. 1950. The identity of the American vulture described as *Cathartes burrovianus* by Cassin. J. Wash. Acad. Sci. 416–418.

Wetmore, A. 1964. A revision of the Americn vultures of the genus *Cathartes*. Smithsonian Misc. Coll. 146:1–18.

Greater Yellow-headed Vulture

Campbell, M. O'N. 2015. Vultures: their evolution, ecology, and conservation. Boca Raton: CRC Press.

Gomez, L. G., D. C. Houston, P. Cotton, and A. Tye. 1995. The role of Greater Yellow-headed Vultures *Cathartes melambrotus* as scavengers in neotropical forest. Ibis 136:193–196.

Graves, G. R. 1992. Greater Yellow-headed Vulture (*Cathartes melambrotus*) locates food by olfaction. J. Raptor Res. 26:38–39.

Houston, D. C. 1994. Observation on Greater Yellow-headed Vulture *Cathartes melambrotus* and other *Cathartes* species as scavengers in forest in Venezuela. *In* Raptor conservation today, ed. B.-U. Meyburg and R. D. Chancellor. London: World Work. Grp. Birds of Prey and Owls.

Mallon, J. M., K. Swing, and D. Mosquera. 2013. Neotropical vulture scavenging succession at a capybara carcass in eastern Ecuador. Ornithol. Neotrop. 24:475–480.

Wetmore, A. 1964. A revision of the Americn vultures of the genus *Cathartes*. Smithsonian Misc. Coll. 146:1–18.

King Vulture

Campbell, M. O'N. 2015. Vultures: their evolution, ecology, and conservation. Boca Raton: CRC Press.

Houston, D. C. 1984. Does the King Vulture *Sarcoramphus papa* use a sense of smell to locate food? Ibis 126:67–69.

Johnson, J. A., J. W. Brown, J. Fuchs, and D. P. Mindell. 2016. Multi-locus phylogenetic inference among New World Vultures (Aves: Cathartidae). Mol. Phylogen. Evol. 105:193–199.

Lemon, W. C. 1991. Foraging behavior of a guild of neotropical vultures. Wilson Bull. 103:698–702.

Mindell, D. J., J. Fuchs, and J. A. Johnson. 2018. Phylogeny, taxonomy, and geographic diversity of diurnal raptors: Falconiformes, Accipitriformes, and Cathartiformes. Birds of prey: biology and conservation in the XXI century, ed. J. H. Sarasola, J. M. Grande, and J. J. Negro. Cham, Switzerland: Springer Nature.

Noriega, J. I., and J. I. Areta. 2005. First record of *Sarcoramphus* Dumeril 1806 (Ciconiiformes: Vulturidae) from the Pleistocene of Buenos Aires Province, Argentina. J. S. Am. Earth Sci. 20:73–79.

Reid, S. B. 1989. Flying behavior and habitat preferences of the King Vulture *Sarcoramphus papa* in the western Orinoco Basin of Venezuela. Ibis 131:301–303.

Schlee, M. A. 1991. Wattle conformation as a criterion for recognizing individual adult King Vultures *Sarcoramphus papa*. Vulture News 25:5–9.

Schlee, M. A. 2000. The status of vultures in Latin America. *In* Raptors at risk, ed. R. D. Chancellor and B.-U. Meyburg. Berlin: World Work. Grp. Birds of Prey and Owls.

Schlee, M. A. 2000. Post-humous presentation of Carl B. Koford's manuscript on a nesting of the King Vulture *Sarcoramphus papa* in Barro Colorado Island, Panama. Vulture News 43:23–29.

Schlee, M. A. 2005. King Vulture (*Sarcoramphus papa*) forage on moriche and cucurit palm stands. J. Raptor Res. 29:269–272.

Schlee, M. A. 2007. King Vultures (*Sarcoramphus papa*) following jaguars in Serrania de la Cerbatana, Venezuela. Vulture News 57:4–16.

Wallace, M. P., and S. A. Temple. 1987. Competitive interactions within and between species in a guild of avian scavengers. Auk 104:290–295.

California Condor

Belding, L. 1890. Land birds of the Pacific district. Occasional papers of the California Academy of Sciences 2. San Francisco: California Acad. Sci.

Chamberlain, C. P., J. R. Waldbauer, K. Fox-Dobbs, S. D. Newsome, P. L. Koch, D. R. Smith, M. E. Church, S. D. Chamberlain, K. J. Sorenson, and R. Riseborough. 2005. Pleistocene to recent dietary shifts in California Condor. Proc. Nat. Acad. Sci. 102:16707–16711.

Cooper, J. G. 1890. A doomed bird. Zoe 1:248–249.
D'Elia, J., and S. M. Haig, eds. 2013. California Condors in the Pacific Northwest. Corvallis: Oregon State University Press.
Geyer, C. J., O. A. Ryder, L. G. Chemnick, and E. A. Thompson. 1993. Analysis of relatedness in the California condors from DNA fingerprints. Mol. Biol. Evol. 10:571–589.
Koford, C. B. 1950. The natural history of the California Condor (*Gymnogyps californianus*). Berkeley: Unpubl. Ph. D. thesis.
Mee, A., and L. S. Hall, eds. 2007. California Condors in the 21st century. Ser. Ornithol 2:1–279.
Rivers, J. W., J. M. Johnson, S. M. Haig, C. J. Schwarz, L. J. Burnett, J. Brandt, D. George, and J. Grantham. 2014. An analysis of monthly home range size in the critically endangered California Condor *Gymnogyps californianus*. Bird Conserv. Int. 24:492–504.
Snyder, N. F. R. 1986. California Condor recovery program. Raptor Res. Rpt. 5:56–71.
Snyder, N., and H. Snyder. 2000. The California Condor: a saga of natural history and conservation. San Diego: Academic Press.
Thwaites, R. G., ed. 1905. Original journals of the Lewis and Clark Expedition, 1804–1806. New York: Dodd Mead.

Andean Condor

Alarcón, P. A. E., J. M. Morales, J. A. Donázar, J. A. Sanchez-Zapata, F. Hiraldo, and S. A. Lambertucci. 2017. Sexual-size dimorphism modulates the trade-off between exploiting food and wind resources in a large avian scavenger. Sci. Rpt. 7:11461.
Alcaide, M., L. Cadahia, S. A. Lambertucci, and J. J. Negro. 2010. Noninvasive estimation of minimum population sizes and variability of the major histocompatibility complex in the Andean Condor. Condor 112:470–478.
Bildstein, K. L. 2004. Raptor migration in the Neotropics: patterns, processes, and consequences. Ornithol. Neotrop. 15(Suppl.):83–99.
Carrete, A., S. A. Lambertucci, K. Speziale, O. Ceballos, A. Travaini, M. Delibes, F. Hiraldo, and J. A. Donazar. 2010. Winners and losers in human-made habitats: interspecific competition outcomes in two Neotropical vultures. Anim. Conserv. 13:390–398.
Darwin, C. 1839. The Voyage of the Beagle. New York: Random House.
Donazar, J. A., A. Travaini, O. Ceballos, A. Rodriguez, M. Delibes, and F. Hiraldo. 1999. Effects of sex-associate competitive asymmetries on foraging group structure and despotic distribution in Andean Condors. Behav. Ecol. Sociobiol. 45:55–65.
Johnson, J. A., J. W. Brown, J. Fuchs, and D. P. Mindell. 2016. Multi-locus phylogenetic inference among New World Vultures (Aves: Cathartidae). Mol. Phylogenet. Evol. 105:193–199.
Lambertucci, S. A., and O. A. Mastrantuoni. 2008. Breeding behavior of a pair of free-living Andean Condors. J. Field Ornithol. 79:147–151.
Lambertucci, S. A., P. A. E. Alarcón, F. Hiraldo, J. A. Sanchez-Zapata, G. Blanco, and J. A. Donázar. 2014. Biol. Conserv. 170:145–150.
Lambertucci, S. A., J. Navarro, J. A. Sanchez Zapata, K. A. Hobson, P. A. E. Alarcón, G. Wiemeyer, G. Blanco, F. Hiraldo, and J. A. Donázar. 2018. Tracking data and retrospective analyses of diet reveal the consequences of loss of marine subsidies for an obligate scavenger, the Andean Condor. Proc. Royal Soc. Lond. B 285:2018.0550.
Lambertucci, S. A., A. Trejo, S. Di Martino, J. A. Sánchez-Zapata, J. A. Donázar, and F. Hiraldo. 2009. Spatial and temporal patterns in the diet of the Andean Condor: ecological replacement of native fauna by exotic species. Anim. Conserv. 12:338–345.

McGahan, J. 1972. Behavior and ecology of Andean Condors, Parts I-III. Madison: Unpubl. Ph.D. diss.
McGahan, J. 1973. Gliding flight of the Andean Condor in nature. J. Exp. Biol. 58:225–237.
McGahan, J. 1973. Flapping flight of the Andean Condor in nature. J. Exp. Biol. 58:239–253.
Mindell, D. J., J. Fuchs, and J. A. Johnson. 2018. Phylogeny, taxonomy, and geographic diversity of diurnal raptors: Falconiformes, Accipitriformes, and Cathartiformes. 2018. *In* Birds of prey: biology and conservation in the XXI century, ed. J. H. Sarasola, J. M. Grande, and J. J. Negro. Cham, Switzerland: Springer Nature.
Padró, J., S. A. Lambertucci, P. L. Perrig, and J. A. Pauli. 2018. Evidence of genetic structure in a wide-ranging and highly mobile soaring scavenger, the Andean condor. Divers. Distrib. 24:1534–1544.
Shepard, E. L. C., S. A. Lambertucci, D. Vallmitjana, and Rory P. Wilson. 2011. Energy beyond food: foraging theory informs time spent in thermals by a large soaring bird. PLoS One 6(11)e27375.
Sick, H. 1993. Birds in Brazil. Princeton: Princeton University Press.
Williams, H. J., E. L. C. Shepard, M. D. Holton, P. A. E. Alarcon, R. P. Wilson, and S. A. Lambertucci. 2020. Physical limits of flight performance in the heaviest soaring bird. Proc. Nat. Acad. Sci.:doi 1907360117.

Palm-nut Vulture

Carneiro, C., M. Henriques, C. Barbosa, Q. Tchantchalam, A. Regalla, A. R. Patrico, and P. Catry. 2017. Ecology and behavior of Palm-nut Vultures *Gypohierax angolensis* in the Bijagós Archipelago, Guinea-Bissau. Ostrich 88:113–121.
Lerner, H. R. L., and D. P. Mindell. 2005. Phylogeny of eagles, Old World vultures, and other Accipitridae based on nuclear and mitochondrial DNA. Mol. Phylog. Evol. 37:327–346.
Mindell, D.P., J. Fuchs, and J. A. Johnson. 2018. Phylogeny, taxonomy, and geographic diversity of diurnal raptors: Falconiformes, Accipitriformes, and Cathartiformes. *In* Birds of prey, ed. J. H. Sarasola, J. M. Grande, and J. J. Negro. Cham, Switzerland: Springer Nature.
Moreau, R. E. 1933. A note on the distribution of the Vulturine Fish-eagle, *Gypohierax angolensis*. Gmel. J. Anim. Ecol. 2:179–183.
Mundy, P., D. Butchart, J. Ledger, and S. Piper. 1992. The vultures of Africa. London: Academic Press.
Thiollay, J.-M. 1978. Les rapaces de une zone de contact savane-forêt en Côte'dIvoire spécialisations alimentaires. Alauda 46:147–170.
Thompson, A. A., and R. E. Moreau. 1957. Feeding habits of the Palm-nut Vulture *Gypohierax*. Ibis 99:608–613.

Bearded Vulture

Delhey, K., A. Peters, and B. Kempenaers. 2007. Cosmetic coloration in birds: occurrence, function, and evolution. American Naturalist 169 (Supplement):S145–S158.
Brown, C. J. 1988. A study of bearded vultures *Gypaetus barbatus* in southern Africa. Pietermaritizburg: Unpubl. Ph.D. diss.
Brown, C. J., and I. Plug. 1990. Food choice and diet of the bearded vulture *Gypaetus barbatus* in southern Africa. S. Afr. J. Zool. 25:169–177.

Brown, C. J., and A. G. Bruxton. 1991. Plumage color and feather structure of the bearded vulture (*Gypaetus barbatus*). J. Zool. 623:627–640.

Delhey, K., A. Peters, and B. Kempenaers. 2007. Cosmetic coloration in birds: occurrence, function, and evolution. Am. Nat. 169 (Supplement):S145–S158.

Fasce, P., and L. Fasce. 2012. First polygynous trio of Bearded Vultures. J. Raptor Res. 46:216–219.

Houston, D. C., A. Hall, H. Frey. 1993. The characteristics of the cosmetic soils use by Bearded Vultures Gypaetus barbatus. Bull. Brit. Ornithol. Cl. 113:260–263.

Houston, D. C., and J. A. Copsey. 1994. Bone digestion intestinal morphology of the Bearded Vulture. J. Raptor Res. 28:73–78.

Krüger, S. 2007. Polyandrous trios in the southern Bearded Vulture *Gypaetus barbatus meridionalis*? Vulture News 57:60–61.

Krüger, S., and A. Amar. 2017. Insights into post-fledging dispersal of Bearded Vultures (*Gypaetus barbatus*) in southern Africa from GPS satellite telemetry. Bird Study 64:125–131.

Margalida, A. 2008. Bearded Vultures (*Gypaetus barbatus*) prefer fatty bones. Behav. Ecol. Sociobiol. 63:187–193.

Margalida, A., and J. Bertran. 2000. Breeding behavior of the Bearded Vulture *Gypaetus barbatus*: minimal sexual difference in parental activities. Ibis 142:225–234.

Margalida, A., and J. Bertran. 2003. Interspecific and intraspecific interaction of Bearded Vulture (*Gypaetus barbatus*) at nesting areas. J. Raptor Res. 37:157–160.

Margalida, A., and J. Bertran. 2005. Territorial defense an agonistic behavior of breeding Bearded Vultures *Gypaetus barbatus* toward conspecifics and heterospecifics. Ethol. Ecol. Evol. 17:51–63.

Margalida, A., J. Bertran, and J. Boudet. 2005. Assessing the diet of nestling Bearded Vultures: a comparison between direct observation methods. J. Field Ornithol. 76:40–45.

Margalida, A., J. Bertran, J. Boudet, and R. Heredia. 2004. Hatching asynchrony, sibling aggression and cannibalism in the Bearded Vulture *Gypaetus barbatus*. Ibis 146:386–393.

Margalida, A., D. Garcia, J. Bertran, and R. Heredia. 2003. Breeding biology and success of the Bearded Vulture *Gypaetus barbatus* in the eastern Pyrenees. Ibis 145:244–252.

Margalida, A., R. Heredia, M. Razin, and M. Hernández. 2008. Sources of variation in mortality of Bearded Vultures *Gypaetus barbatus* in Europe. Bird Conserv. Int. 18:1–10.

Mundy, P., D. Butchart, J. Ledger, and S. Piper. 1992. The vultures of Africa. London: Academic Press.

Negro, J. J., A. Margalida, F. Hiraldo, and R. Heredia. 1999. The function of the cosmetic coloration of bearded vultures: when art imitates life. Anim. Behav. 58(Forum):F14–F17.

van Overveld, T., M. de la Riva, J. A. Donázar. 2017. Cosmetic coloration in Egyptian vultures: mud bathing as a tool for social communication. Ecol. 98:2216–2218.

Xirouchakis, S. 1998. Dust bathing in the Bearded Vulture (*Gypaetus barbatus*). J. Raptor Res. 32:322.

Egyptian Vulture

Agudo, R., C. Rico, C. Vilà, F. Hiraldo, and J. A. Donázar. 2010. The role of humans in the diversification of a threatened island raptor. BMC Evol. Biol. 10:384-392.

Al Fazari, W., and M. J. McGrady. 2016. Counts of Egyptian Vultures *Neophron percnopterus* and other avian scavengers at Muscat's municipal landfill, Oman, November 2013–March 2015. Sandgrouse 38:99–105.

Andersson, J. 1970. Lake Ngami or explorations and discovery during four years of wanderings in the wilds of south western Africa. London: Hurst and Beckett.

Arkumarev, V., V. Dobrev, Y. D. Abebe, G. Popogeorgiev, and S. C. Nikolov. 2014. Congregations of wintering Egyptian Vultures *Neophron percnopterus* in Afar, Ethiopia: present status an implication for conservation. Ostrich 85:139–145.

Barcell, M., J. R. Benítez, F. Solera, B. Román, and J. A. Donázar. 2015. Egyptian Vulture (*Neophron percnopterus*) uses stone-throwing to break into a Griffon Vulture (*Gyps fulvus*) egg. J. Raptor Res. 49:521–522.

Buechley, M., J. McGrady, E. Çoban, and C. H. Şekercioğlu. 2018. Satellite tracking a wide-ranging endangered vulture species to target conservation actions in the Middle East and Africa. Biodivers. Conserv. 27:2293–2310.

Carrete, M., A. Centeno-Cuadros, M. Méndez, R. Agudo, and J. A. Donázar. 2017. Low heritability in tool use in a wild vulture population. Anim. Behav. 129:127–131.

Ceballos, O., and J. A. Donázar. 1990. Roost-tree characteristics, food habitats and season abundance of roosting Egyptian Vultures in northern Spain. J. Raptor Res. 24:19–25.

Donázar, J. A., and O. Cabellos. 1989. Growth rates of nestling Egyptian Vulture *Neophron percnopterus* in relation to brood size, hatching order and environmental factors. Ardea 77:217–226.

Donázar, J. A., and O. Cabellos. 1990. Post-fledging dependence period and development of flight and foraging behavior in the Egyptian Vulture *Neophron percnopterus*. Ardea 78:387–394.

Donázar, J. A., J. J. Negro, C. J. Palacios, L. Gangoso, J. A. Godoy, O. Cabellos, F. Hiraldo, and N. Capote. 2002. Description of a new subspecies of the Egyptian Vulture (Accipitridae: *Neophron percnopterus*) from the Canary Islands. J. Raptor Res. 36:17–23.

Garcia, E. F. J., and K. J. Bensusan. 2006. Northbound migrant raptors in June and July at the Strait of Gibraltar. Brit. Birds 99:569–575.

García-Ripollés, C., and P. Lopéz-Lopéz. 2006. Population size and breeding performance of Egyptian Vultures (*Neophron percnopterus*) in eastern Iberian Peninsula. J. Raptor Res. 40:217–221.

García-Ripollés, C., P. Lopéz-Lopéz, and V. Urios. 2010. First description of migration and wintering of adult Egyptian Vultures *Neophron percnopterus* tracked by GPS satellite telemetry. Bird Study 57:261–265.

McGrady, M. J., D. L. Karelus, H. A. Rayaleh, M. S. Willson, B.-U. Meyburg, M. K. Oli, and K. Bildstein. 2018. Home ranges and movements of Egyptian Vultures *Neophron percnopterus* in relation to rubbish dumps in Oman and the Horn of Africa. Bird Study 65:544–556.

Meyburg, B.-U., M. Gallardo, C. Meyburg, and E. Dimitrova. 2004. Migrations and sojourn in Africa of Egyptian vultures (*Neophron percnopterus*) tracked by satellite. J. Ornithol. 145:273–280.

Mundy, P., D. Butchart, J. Ledger, and S. Piper. 1992. The vultures of Africa. London: Academic Press.

Negro, J. J., J. M. Grande, J. L. Tella, J. Garrido, D. Hornero, J. A. Donázar, J. A. Zapata, J. R. Benítez, and M. Barcell. 2002. Nature 416:807–808.

Stoyanova, Y., N. Stefanov, and J. K. Schmutz. 2010. Twig used as tool by the Egyptian Vulture (*Neophron percnopterus*). J. Raptor Res. 44:154–156.

Thouless, C. R., J. H. Fanshawe, and B. C. R. Bertram. 1989. Egyptian Vultures *Neophron percnopterus* and Ostrich *Struthio camelus* eggs: the origins of stone-throwing behavior. Ibis 131:9–15.

van Lawick-Goodall, J., and H. van Lawick. 1968. The use of tools by Egyptian Vulture *Neophron percnopterus*. Nature 212:290–294.

Hooded Vulture

Dabone, C., R. Buij, A. Oued, J. B. Adajakpa, W. Wendengoudi, and P. D. M. Weesie. 2019. Impact of human activities on the reproduction of Hooded Vultures *Necrosyrtes monachus* in Burkina Faso. Ostrich 90:53–61.
Jallow, M., C. R. Barlow, L. Sanyang, L. Dibba, C. Kendall, M. Bechard, and K. L. Bildstein. 2016. High population of the critically endangered Hooded Vulture *Necrosyrtes monachus* in Western Region, The Gambia, confirmed by road surveys in 2013 and 2015. Malimbus 38:23–28.
Monadjem, A., K. Wolter, W. Neser, and K. L. Bildstein. 2016. Hooded Vulture *Necrosyrtes monachus* and African White-backed Vulture *Gyps africanus* nesting at the Olifants River Private Nature Reserve, Limpopo Province, South Africa. Ostrich 87:113–117.
Mundy, P., D. Butchart, J. Ledger, and S. Piper. 1992. The vultures of Africa. London: Academic Press.
Ogada, D. L., and R. Buij. 2011. Large declines of the Hooded Vulture *Necrosyrtes monachus* across its African range. Ostrich 82:101–113.
Pomeroy, D. E. 1975. Birds as scavengers of refuse in Uganda. Ibis 117:69–81.
Reading, R. P., B. Tshimologo, and G. Maude. 2017. Coprophagy of African Wild Dog faeces by Hooded Vultures in Botswana. Vulture News 72:34–37.
Roche, C. 2006. Breeding records and nest site preference of Hooded Vultures in greater Kruger National Park. Ostrich 77:99–101.
Thompson, L. J., J. P. Davies, K. L. Bildstein, and C. T. Downs. 2017. Removal (and attempted removal) of nest material from a Hooded Vulture *Necrosyrtes monachus* nest by a starling and a Hooded Crow. Ostrich 88:183–187.
Thompson, L. J., J. P. Davies, M. Gudehus, A. J. Botha, K. L. Bildstein, C. Murn, and C. T. Downs. 2017. Visitors to nests of Hooded Vultures *Necrosyrtes monachus* in north-eastern South Africa. Ostrich 88:155–162.
Thompson, L. J., D. Barber, M. Bechard, A. J. Botha, K. Wolter, W. Neser, E. R. Buechley, R. Reading, R. Garbett, P. Hancock, G. Maude, M. Z. Virani, S. Thomsett, H. Lee, D. Ogada, C. Barlow, and K. L. Bildstein. 2020. Variation in monthly home ranges of Hooded Vultures *Necrosyrtes monachus* in western, eastern, and southern Africa. Ibis. doi 10.1111/bi.12836.
Van Someren, V. G. L. 1956. Days with birds: studies of habits of some East African species. Fieldiana (Zool.) 38:38–47.

Indian (aka Long-billed) Vulture

Ali, S., and S. Dillon Ripley. 1978. Handbook of the birds of India and Pakistan. Second edition. London: Oxford University Press.
Arshad, M., J. Gonzalez, A. A. El-Sayed, T. Osborne, and M. Wink. 2009. Phylogeny and phylogeography of critically endangered *Gyps* species based on nuclear and mitochondrial markers. J. Ornithol. 150:419–430.
Grubh, R. B. 1978. Competition and co-existence in griffon vultures: *Gyps bengalensis*, *G. indicus* & *G. fulvus* in the Gir Forest. J. Bombay Nat. Hist. Soc. 75:810–814.

Grubh, R. B. 1978. The griffon vultures (*Gyps bengalensis, G. indicus,* and *G. fulvus*) of the Gir Forest: their feeding habitats and the nature of association with the Asiatic lion. J. Bombay Nat. Hist. Soc. 75:1058–1068.
Johnson, J. A., H. R. L. Lerner, P. C. Rasmussen, and D. P. Mindell. 2006. Systematics within *Gyps* vultures: a clade at risk. BMC Evol. Biol. 6:65–75.
Livesey, T. R. 1939. Vultures feeding at night. J. Bombay Nat. Hist. Soc. 40:755–756.
Naoroji, R. 2006. Birds of prey of the Indian subcontinent. London: Christopher Helm.

Slender-billed Vulture

Ali, S., and S. Dillon Ripley. 1978. Handbook of the birds of India and Pakistan. Second edition. London: Oxford University Press.
Arshad, M., J. Gonzalez, A. A. El-Sayed, T. Osborne, and M. Wink. 2009. Phylogeny and phylogeography of critically endangered *Gyps* species based on nuclear and mitochondrial markers. J. Ornithol. 150:419–430.
Johnson, J. A., H. R. L. Lerner, P. C. Rasmussen, and D. P. Mindell. 2006. Systematics within *Gyps* vultures: a clade at risk. BMC Evol. Biol. 6:65–75.
Livesey, T. R. 1939. Vultures feeding at night. J. Bombay Nat. Hist. Soc. 40:755–756.
Naoroji, R. 2006. Birds of prey of the Indian subcontinent. London: Christopher Helm.
Rasmussen, P. C., and S. J. Parry. 2001. Taxonomic status of the "Long-billed" Vulture *Gyps indicus*. Vulture News 44:18–21.
Rasmussen, P. C., W. S. Clark, S. J. Parry, and J. Schmitt. 2001. Field identification of 'Long-billed' Vultures (Indian and Slender-billed Vultures) Oriental Bird Cl. Bull. 34:24–29.

White-rumped Vulture

Ali, S., and S. Dillon Ripley. 1978. Handbook of the birds of India and Pakistan. Second edition. London: Oxford University Press.
Arshad, M., J. Gonzalez, A. A. El-Sayed, T. Osborne, and M. Wink. 2009. Phylogeny and phylogeography of critically endangered *Gyps* species based on nuclear and mitochondrial markers. J. Ornithol. 150:419–430.
Baral, N., R. Gautam, and B. Tamang. 2005. Population status and breeding ecology of White-rumped Vulture *Gyps bengalensis* in Rampur Valley, Nepal. Forktail 21:87–91.
Galushin, V. M. 1971. A huge urban population of birds of prey in Delhi, India. Ibis 113:522.
Grubh, R. B. 1978. Competition and co-existence in griffon vultures: *Gyps bengalensis, G. indicus,* and *G. fulvus* in the Gir Forest. J. Bombay Nat. Hist. Soc. 75:810–814.
Grubh, R. B. 1978. The griffon vultures (*Gyps bengalensis, G. indicus,* and G. *fulvus*) of the Gir Forest: their feeding habitats and the nature of association with the Asiatic lion. J. Bombay Nat. Hist. Soc. 75:1058–1068.
Johnson, J. A., H. R. L. Lerner, P. C. Rasmussen, and D. P. Mindell. 2006. Systematics within *Gyps* vultures: a clade at risk. BMC Evol. Biol. 6:65–75.
Livesey, T. R. 1939. Vultures feeding at night. J. Bombay Nat. Hist Soc. 40:755–756.
Naoroji, R. 2006. Birds of prey of the Indian subcontinent. London: Christopher Helm.
Prakash, V. 1999. Status of vultures in Keoladeo National Park, Bharaptur, Rajasthan, with special reference to population crash in *Gyps* species. J. Bombay Nat. Hist. Soc. 96:365–378.
Thakur, M. L. 2015. Breeding ecology and distribution of White-rumped Vultures (*Gyps bengalensis*) in Himachal Pradesh, India. J. Raptor Res. 49:183–191.

Griffon Vulture

Ali, S., and S. Dillon Ripley. 1978. Handbook of the birds of India and Pakistan. Second edition. London: Oxford University Press.
Anonymous. 2008. Predator vultures eat Spanish farm animals alive. Vulture News 58:50–51.
Arshad, M., J. Gonzalez, A. A. El-Sayed, T. Osborne, and M. Wink. 2009. Phylogeny and phylogeography of critically endangered *Gyps* species based on nuclear and mitochondrial markers. J. Ornithol. 150:419–430.
Bildstein, K. L., M. J. Bechard, C. Farmer, and L. Newcomb. 2009. Narrow sea crossings present major obstacles to migrating Griffon Vultures *Gyps fulvus*. Ibis 151:382–391.
Blanco, G., and F. Martinez. 1996. Sex difference in breeding age of Griffon Vultures (*Gyps fulvus*). Auk 113:247–248.
Fernández, C., and J. A. Donàzar. 1991. Griffon Vultures *Gyps fulvus* occupying eyries of other cliff-nesting raptors. Bird Study 38:42–44.
García-Ripollés, C., P. López-López, and V. Urios. 2011. Ranging behavior of non-breeding Eurasian Griffon Vultures *Gyps fulvus*: a GPS-telemetry study. Acta Ornithol. 46:127–134.
Goodwin, D. 1949. Notes on the migration of birds of prey over Suez, 1947. Ibis 91:59–63.
Green, R., J. A. Donázar, J. Sánchez-Zapata, and A. Margalida. 2016. Potential threat to Eurasian griffon vultures in Spain from veterinary use of the drug diclofenac. J. Appl. Ecol. 53:993–1003.
Grubh, R. B. 1978. Competition and co-existence in Griffon Vultures: *Gyps bengalensis*, *G. indicus*, and *G. fulvus* in the Gir Forest. J. Bombay Nat. Hist. Soc. 75:810–814.
Grubh, R. B. 1978. The Griffon Vultures (*Gyps bengalensis*, *G. indicus*, and *G. fulvus*) of the Gir Forest: their feeding habitats and the nature of association with the Asiatic lion. J. Bombay Nat. Hist. Soc. 75:1058–1068.
Johnson, J. A., H. R. L. Lerner, P. C. Rasmussen, and D. P. Mindell. 2006. Systematics within *Gyps* vultures: a clade at risk. BMC Evol. Biol. 6:65.
Naoroji, R. 2006. Birds of prey of the Indian subcontinent. London: Christopher Helm.
Parra, J., and J. L. Tellería. 2004. The increase in the Spanish population of Griffon Vulture *Gyps fulvus* during 1989–1999: effects of food and nest-site availability. Bird Conserv. 14:33–41.
Pitches, A. 2016. Diclofenac threat to Spanish vultures. Br. Birds 109:312.
Seguí, A., A. Belda, P. M. Mojoica, and M. B. Zaragozi. 2018. Effectiveness of the use of patagial wing tags for griffon vultures (*Gyps fulvus*) in Spain. Arxius Misc. Zool. 16:255–270.
Slotta-Bachmyrl, L., R. Bögel, and C. Cardenal A., eds. 2004. The Eurasian Griffon Vulture (*Gyps fulvus*) in Europe and the Mediterranean; status report and action plan. East Salzburg: Eur./Mediterranean Griffon Vulture Act. Grp.
Susic, G. 2000. Regular long-distance migration of Eurasian Griffon *Gyps fulvus*. *In* Raptors at risk, ed. R. D. Chancellor and B.-U. Meyburg. Berlin: World Work. Grp. Birds of Prey and Owls.
Xirouchakis, S. M. 2010. Breeding biology and reproductive performance of Griffon Vultures *Gyps fulvus* on the island of Crete (Greece). Bird Study 57:213–225.

Rüppell's Vulture

Hiebl, I., R. E. Weber, D. Schneeganss, J. Kosters, and G. Braunitzer. 1988. Structural adaptation in the major and minor hemoglobin components of adult Rüppells Griffon (*Gyps rueppellii*): a new molecular pattern for hypoxia tolerance. Biol. Chem. Hoppe-Seyler 369:217–232.

Houston, D. C. 1974. The role of Griffon Vultures *Gyps africanus* and *Gyps rueppellii* as scavengers. J. Zool. Lond. 172:35–46.

Houston, D. C. 1976. Breeding of White-backed and Rueppell's Griffon Vultures, *Gyps africanus* and *G. rueppellii*. Ibis 118:14–40.

Houston, D. C. 1983. The adaptive radiation of Griffon Vultures. *In* Vulture biology and management, ed. S. R. Wilbur and J. A. Jackson. Berkeley: University of California Press.

Johnson, J. A., H. R. L. Lerner, P. C. Rasmussen, and D. P. Mindell. 2006. Systematics within *Gyps* vultures: a clade at risk. BMC Evol. Biol. 6:65–75.

Kendall, C. J., M. Z. Virani, J. G. C. Hopcraft, K. L. Bildstein, and D. I. Rubenstein. 2014. African vultures don't follow migratory herds: scavenger habitat use is not mediated by prey abundance. PLoS ONE 9:e83470.

Kruuk, H. 1967. Competition for food between vultures in East Africa. Ardea 55:171–193.

Laybourne, R. C. 1974. Collision between a vulture and an aircraft at an altitude of 37,000 feet. Wilson Bull. 86:461–462.

Ramirez, J., A. R. Muñoz, A. Onrubia, A. de la Cruz, D. Cuenca, J. M. González, and G. M. Arroyo. 2011. Spring movements of Rüppell's Vulture *Gyps ruepellii* across the Strait of Gibraltar. Ostrich 82:71–73.

Rodríguez, G., and J. Elorriaga. 2016. Identification of Rüppell's Vulture and White-backed Vulture and vagrancy in the WP. Dutch Bird. 38:349–375.

Román, J. R. 2012. First record of Rüppells Vulture *Gyps rueppellii* arriving in Morocco from Spain. Go-South Bull. 9:44–45.

Rondeau, G., P. Pilard, B. Ahon, and M. Condé. 2006. Tree nesting Ruppell's Griffon Vultures. Vulture News 55:1–10.

Thévenot, M., R. Vernon, and P. Bergier. 2003. The birds of Morocco. Tring: British Ornithological Union.

Virani, M. Z., A. Monadjem, S. Thomsett, and C. Kendall. 2012. Seasonal variation in breeding Rüppell's Vultures *Gyps ruepellii* at Kwenia, southern Kenya, and implications for conservation. Bird Conserv. Int. 22:260–269.

Himalayan Vulture

Acharya, R., R. Cuthbert, H. S. Baral, and K. B. Shah. 2009. Rapid population declines of Himalayan Griffons *Gyps himalayensis* in Upper Mustang, Nepal. Bird Conserv. Int. 19:99–107.

Johnson, J. A., H. R. L. Lerner, P. C. Rasmussen, and D. P. Mindell. 2006. Systematics within *Gyps* vultures: a clade at risk. BMC Evol. Biol. 6:65.

Li, Y. D., and C. Kasorndorkbua. 2008. The status of the Himalayan Griffon *Gyps himalayensis* in South-East Asia. Forktail 24:57–62.

Lu, X., D. Ke, X. Zeng, G. Gong, and R. Ci. 2009. Status, ecology, and conservation of the Himalayan Griffon *Gyps himalayensis* (Aves, Accipitridae) in the Tibetan Plateau. Ambio 38:166–173.

Naoroji, R. 2006. Birds of prey of the Indian subcontinent. London: Christopher Helm.

Praveen, J., P. O. Nameer, D. Karuthedathu, C. Ramaiah, B. Balakrishnan, K. M. Rao, S. Shurpali, R. Puttaswamaiah, and I. Tavcar. 2014. On the vagrancy of the Himalayan Vulture *Gyps himalayensis* to southern India. Indian Birds 9:19–22.

Sherub, S. 2017. Movement mechanisms of *Gyps himalayensis* (Himalayan Vultures) in the Central Asian Flyway. Konstanz: Unpubl. Ph.D. diss.

Virani, M. Z., J. B. Giri, R. T. Watson, and H. S. Baral. 2008. Surveys of Himalayan Vultures (*Gyps himalayensis*) on the Annapurna Conservation Area, Mustang, Nepal. J. Raptor Res. 42:197–203.

White-backed Vulture

Kendall, C. J., M. Z. Virani, J. G. C. Hopcraft, K. L. Bildstein, and D. I. Rubenstein. 2014. African vultures don't follow migratory herds: scavenger habitat use is not mediated by prey abundance. PLoS ONE 9:e83470.

Monadjem, A., K. Wolter, W. Neser, and K. Bildstein. 2016. Hooded Vulture *Necrosyrtes monachus* and African White-backed Vulture *Gyps africanus* nesting at the Olifants River Private Reserve, Limpopo Province, South Africa. Ostrich 87:113–117.

Mundy, P., D. Butchart, J. Ledger, and S. Piper. 1992. The vultures of Africa. London: Academic Press.

Phipps, W. L., S. G. Willis, K. Wolter, and V. Naidoo. 2013. Foraging ranges of immature African White-backed Vultures (*Gyps africanus*) and their use of protected areas in southern Africa. PLoS ONE 8:e52813.

Virani, M., P. Kirui, A. Monadjem, S. Thomsett, and M. Githiru. 2010. Nesting status of African White-backed Vultures *Gyps africanus* in the Masai Mara National Reserve, Kenya. Ostrich 81:205–209.

Cape Vulture

Bamford, A. J., M. Diekmann, A. Monadjem, and J. Mendelsohn. 2007. Ranging behavior of Cape Vulture *Gyps coprotheres* from an endangered population in Namibia. Bird Conserv. Int. 17:331–339.

Boshoff, A., and M. D. Anderson. 2007. Towards a conservation plan for the Cape Griffon *Gyps coprotheres*: identifying priorities for research and conservation. Vulture News 57:56–59.

Boshoff, A., A. Barkhuysen, G. Brown, and M. Michael. 2009. Evidence of partial migratory behavior by the Cape Griffon *Gyps coprotheres*. Ostrich 80:129–133.

Hirschauer, M. T., and K. Wolter. 2017. High occurrence of extra-pair partnerships and homosexuality in a captive Cape Vulture *Gyps coprotheres* colony. Ostrich 88:173–176.

Hirschauer, M. T., K. Wolter, and W. Neser. 2017. Natal philopatry in young Cape Vultures *Gyps coprotheres*. Ostrich 88:79–82.

Martens, F. R., M. B. Pfeiffer, C. T. Downs, and J. Q. Venter. 2018. Post-fledging movement and spatial ecology of the endangered Cape Vulture (*Gyps coprotheres*). J. Ornithol. 159:913–922.

Mundy, P., D. Butchart, J. Ledger, and S. Piper. 1992. The vultures of Africa. London: Academic Press.

Pfeiffer, M. B., J. A. Venter, and C. T. Downs. 2016. Cliff characteristics, neighbor requirement and breeding success of the colonial Cape Vulture *Gyps coprotheres*. Ibis 159:26–37.

Phipps, W. L., K. Wolter, M. D. Michael, Lynne M. MacTavish, and R. W. Yarnell. 2013. Do powerline and protected areas present a catch-22 situation for Cape Vultures (*Gyps coprotheres*)? PLoS ONE 8:e76794.

Robertson, A. S. 1985. Observations on the post-fledging dependence period of Cape Vultures. Ostrich 56:58–66.

Robertson, A. 1986. Notes on the breeding cycle of Cape Vultures (*Gyps coprotheres*). Raptor Res. 20:51–60.

Robertson, A. 1986. Copulations throughout breeding in a colonial accipitrid vulture. Condor 88:535–539.

Wolter, K., W. Neser, M. T. Hirschauer, and A. Camaro. 2016. Cape Vulture *Gyps coprotheres* breeding status in southern Africa: monitoring results from 2010–1014. Ostrich 87:119–123.

Red-headed Vulture

Changani, A. K. 2007. Sightings and nesting sites of Red-headed Vulture *Sarcogyps calvus* in Rajasthan, India. Indian Birds 3:218–221.

Dhakal, H., K. M. Baral, K. P. Bhusal, and H. P. Sharma. 2014. First records of nests and breeding success of Red-headed Vulture *Sarcogyps calvus* and implementation of vulture conservation programs in Nepal. Ela J. 3:3–15.

Galligan, T. H., T. Amano, V. M. Prakash, M. Kulkarni, R. Shringarpure, N. Prakash, S. Ranade, R. E. Green, and R. J. Cuthbert. 2014. Have population declines in Egyptian Vulture and Red-headed Vulture in India declined since 2006 ban on veterinary diclofenac? Bird Conserv. Int. 24:272–281.

Hille, S. M., F. Korner-Nievergelt, M. Bleeker, and N. J. Collar. 2016. Foraging behaviour at carcasses in an Asian vulture assemblage: towards a good restaurant guide. Bird Conserv. Int. 26:263–272.

Kumar, A., A. Sinha, and A. Kanaujia. 2017. Sighting of Red-headed Vultures (*Sarcogyps calvus*) in a group. Vulture News 73:13–17.

White-headed Vulture

Amadon, D. 1977. Notes on the taxonomy of vultures. Condor 79:413–416.

Mundy, P., D. Butchart, J. Ledger, and S. Piper. 1992. The vultures of Africa. London: Academic.

Murn, C. 2012. Field identification of individual White-headed Vultures *Trigonoceps occipitalis* using plumage patterns—an information theoretic approach. Bird Study 59:515–521.

Murn, C., and G. J. Holloway. 2014. Breeding biology of the White-headed Vulture *Trigonoceps occipitalis* in Kruger National Park, South Africa. Ostrich 85:125–130.

Murn, C., P. Mundy, M. Z. Virani, W. D. Borello, G. J. Holloway, and J.-M. Thiollay. 2016. Using Africa's protected areas network to estimate the global population of a threatened and declining species: a case study of the Critically Endangered White-headed Vulture *Trigonoceps occipitalis*. Ecol. Evol. 6:1092–1103.

Portugal, S. J., C. P. Murn, and G. R. Martin. 2017. White-headed Vulture *Trigonoceps occipitalis* shows visual field characteristics of hunting raptors. Ibis 159:463–466.

White, C. M. N. 1951. Systematic notes on African birds. Ostrich 22:25–26.

Cinereous Vulture

Batbayar, N., R. Reding, D. Kenny, T. Natsagdorf, and P. W. Kee. 2000. Migration and movement patterns of Cinereous Vultures in Mongolia. Falco 32:5–7.

Castaño, J. P., J. F. Sánchez, M. A. Díaz-Portero, and M. Robles. 2015. Dispersal and survival of juvenile Black Vultures *Aegypius monachus* in central Spain. Ardeola 62:351–361.

Costillo, E., C. Corbacho, R. Morán, and A. Villegas. 2007. The diet of the Black Vulture *Aegypius monachus* in response to environmental changes in Extremadura (1970–2000). Ardeola 54:197–254.

Gavashelishvili, A., M. McGrady, M. Ghasabian, and K. L. Bildstein. 2012. Movements and habitat use by immature Cinereous Vultures (*Aegypius monachus*) from the Caucasus. Bird Study 59:449–462.

Hiraldo, F. 1976. Diet of the Black Vulture (*Aegypius monachus*) in the Iberian Peninsula. Doñana Acta Vertebrata 3:19–31.

Jensen, A. E., T. H. Fisher, R. O. Hutchinson. 2015. Notable new records from the Philippines. Forktail 31:24–36.

Kenny D., N. Batbayar, P. Tsolmonjav, J. Willis, J. Azua, and R. Reading. 2008. Dispersal of Eurasian Black Vulture *Aegypius monachus* fledglings from the Ikh Nart Nature Reserve, Mongolia. Vulture News 59:13–19.

Kim, J.-H., O.-S. Chung, W.-S. Lee, and Y. Kanai. 2007. Migration of Cinereous Vultures (*Aegypius monachus*) in Northeast Asia. J. Raptor Res. 41:161–165.

Mundy, P., D. Butchart, J. Ledger, and S. Piper. 1992. The vultures of Africa. London: Academic Press.

Naoroji, R. 2006. Birds of prey of the Indian subcontinent. London: Christopher Helm.

Reading, R., S. Amgalanbaatar, D. Kenny, and B. Dashdemberel. 2005. Cinereous Vulture ecology in Kkh Nartyn Chuluu Nature Reserve, Mongolia. Mongolian J. Biol. Sci. 3:13–19.

Vroege, J. A. 2013. Raptors from the West-Palearctic observed in The Gambia, Senegal and northern Mauritania in February–March 2005. Takkeling 21:83–87.

Weber, R. E., I. Hiebl, and G. Braunitzer. 1988. High altitude hemoglobin function in the vultures *Gyps rueppellii* and *Aegypius monachus*. Biol. Chem. Hoppe-Seyler 369:233–240.

Yamaç, E., and C. C. Bilgin. 2012. Post-fledging movements of Cinereous Vulture *Aegypius monachus* in Turkey revealed by GPS telemetry. Ardea 100:149–156.

Yamaç, E., and E. Günyel. 2010. Diet of the Eurasian Black Vulture, *Aegypius monachus* Linnaeus, 1766, in Turkey and implications for its conservation. Zool. Middle East 51:15–22.

Lappet-faced Vulture

Bamford, A. J., A. Monadjem, M. D. Anderson, A. Anthony, W. D. Borello, M. Bridgeford, P. Bridgeford, P. Hancock, B. Howells, J. Wakelin, and I. C. W. Hardy. 2009. Tradeoffs between specificity and regional generality in habitat association models: a case study of two species of African vultures. J. Appl. Ecol. 46:852–860.

Garbett, R. A. 2018. Conservation of raptors and vultures in Botswana: with a focus on Lappet-faced Vulture *Torgos tracheliotos*. Capetown: Unpubl. Ph.D. diss.

Kendall, C. J., M. Z. Virani, G. C. Hopcraft, K. L. Bildstein, and D. I. Rubenstein. 2014. African vultures don't follow migratory herds: scavenger habitat use is not mediated by prey abundance. PLoS ONE 9:e83470.

McCulloch G. 2006. Lappet-faced Vultures—social hunters? Vulture News 55:10–13.

Monadjem, A., and D. K. Garcelon. 2005. Nesting distribution of vultures in relation to land use in Swaziland. Biodiver. Conserv. 14:2079–2092.

Newton, S. F., and A. V. Newton. 1996. Breeding biology and seasonal abundance of Lappet-faced Vultures *Torgos tracheliotus* in western Saudi Arabia. Ibis 138:675–683.

Sauer E. G. F. 1973. Notes on the behavior of Lappet-faced Vultures and Cape Vultures in the Namib Desert of South West Africa. Madoqua series II 7:63–68.

Shobrak, M. 1996. Ecology of Lappet-faced Vulture *Torgos tracheliotus* in Saudi Arabia. Glasgow: Unpubl. Ph.D. diss.

Shobrak, M. 2001. Posturing behavior of the Lappet-faced Vulture *Torgos tracheliotus* chicks on the nest plays a role in protecting them from high ambient temperatures. Avian Raptors Bull. 2:7–9.

Shobrak, M. 2011. Changes in the number of breeding pairs, nest distribution and nesting trees used by Lappet-faced Vultures *Torgos tracheliotus* in the Mahazat As-Sayd Protect Area, Saudi Arabia. J. Bombay Nat. Hist. Soc. 108:114–119.

Shobrak, M. 2014. Satellite tracking of the Lappet-faced Vulture *Torgos tracheliotos* in Saudi Arabia. Jordan J. Nat. Hist. 1:131–141.

Chapter 3: Pair Formation and Reproduction

Anto, R. J., A. Margalida, H. Frey, R. Heredia, L. Lorente, and J. A. Sesé. 2007. First breeding in captive and wild Bearded Vulture *Gypaetus barbatus*. Acta Onithol. 42:115–118.

Ardrey, R. 1966. The territorial imperative. New York: Atheneun.

Bertran, J., and A. Margalida. 1999. Copulatory behavior of the Bearded Vulture. Condor 101:164–168.

Bertran, J., and A. Margalida. 2002. Territorial behavior of Bearded Vultures in response to Griffon Vultures. J. Field Ornithol. 73:86–90.

Bildstein, K. L. 1992. Causes and consquences of reversed size dimporphism in raptors: the head-start hypothesis. J. Raptor Res. 26:115–123.

Bildstein, K. L. 2017. Raptors: the curious nature of diurnal birds of prey. Ithaca: Cornell University Press.

Brown, C. J. 1990. Breeding biology of the Bearded Vulture in southern Africa, Part 1. The prelaying and incubation periods. Ostrich 61:24–32.

Darwin, C. 1859. On the origin of species. London: Murray.

Donázar, J. A., O. Ceballos, and J. L. Tella. 1994. Copulation behavior in Egyptian Vulture *Neophron percnopterus*. Bird Study 41:37–41.

Ferrer, M., K. Bildstein, V. Penteriani, E. Casado, and M. de Lucas. 2011. Why birds with deferred sexual maturity are sedentary on islands. PLoS ONE 6:e22056.

Garbett, R. A. 2018. Conservation of raptors and vultures in Botswana: with a focus on Lappet-faced Vultures *Torgos trachliotus*. Cape Town: Unpubl. Ph.D. diss.

Howard, E. 1920. Territory in bird life. London: Howard Murray.

Huber, B. 1939. Siebrohrensystem unserer Baume und seine jahreszeitlichen Veranderungen. Jahrbücher für Wissenschaftliche Botanik. 88:176–242.

Le Gouar, P., J. Sulawa, S. Heniquet, C. Tessier, and F. Sarrazin. 2011. Low evidence of extra-pair fertilizations in two reintroduced populations of Griffon Vulture (*Gyps fulvus*). J. Ornithol. 152:359–364.

López-López, P., J. A. Gil, and M. Alcátara. 2011. Morphometrics and sex determination in the endangered Bearded Vulture. J. Raptor Res. 45:361–365.

Margalida, A., J. Bertran, J. Boudet, and R. Heredia. 2004. Hatching asynchrony, sibling aggression and cannibalism in the Bearded Vultures *Gypaetus barbatus*. Ibis 146:386–393.

Mee, A., G. Ausin, M. Barth, C. Beestman, T. Smith, and M. Wallace. 2004. Courtship behavior in reintroduced California Condors: evidence for extra-pair copulations and female

mate guarding. *In* Raptors worldwide, ed. R. D. Chancellor and B.-U. Meyburg. Berlin: World Work. Grp. Birds of Prey and Owls.
Mundy, P., D. Butchart, J. Ledger, and S. Piper. 1992. The vultures of Africa. London: Academic Press.
Murn, C. 2019. Talon-grappling and cartwheeling of Hooded Vultures in South Africa. J. Raptor Res. 53:353–354.
Negro, J. J., and J. M. Grande. 2001. Territorial signalling: a new hypothesis to explain frequent copulation in raptorial birds. Anim. Behav. 62:803–809.
Newton, I. 1986. The Sparrowhawk. Calton: T & A D Poyser.
Robertson, A. 1986. Copulations throughout breeding in a colonial vulture. Condor 88:535–539.
Rollack, C. E., K. Wiebe, M. Stoffel, and C. S. Houston. 2013. Turkey Vulture breeding behavior studied with trail cameras. J. Raptor Res. 47:153–160.
Sarrazin, F., C. Bagnolini, J. L. Pinna, and E. Danchin. 1996. Breeding biology during establishment of a reintroduced Griffon Vulture *Gyps fulvus* population. Ibis 138:325–325.

Chapter 4: Food Finding and Feeding Behavior

Anderson, D. J., and R. J. Horwitz. 1979. Competitive interactions among vultures and their avian competitors. Ibis 121:505–508.
Bagg, A. M., and H. M. Parker. 1951. The Turkey Vulture in New England and eastern Canada up to 1950. Auk 68:315–338.
Ballejo, F., S. A. Lambertucci, A. Trejo, and L. M. De Santis. 2018. Trophic niche overlap among scavengers in Patagonia supports the condor-vulture competition hypothesis. Bird Conserv. Internat. 28:390–402.
Bank, M. S., and W. L. Franklin. 1998. Puma (*Felis concolor*) feeding observations and attacks on guanacos (*Lama guanicoe*). Mammalia 64:599–605.
Blazquez, M. C., M. Delibes-Mateos, J. M. Vargas, A. Granados, A. Delgado, and M. Delibes. 2016. Stable isotope evidence for Turkey Vulture reliance on food subsidies from the sea. Ecol. Indicat. 63:332–336.
Camiña, A., R. Palomo, and J. C. Torres. 2002. Observation of cannibalism in Eurasian Griffons *Gyps fulvus* in Spain. Vulture News 47:25–26.
Chamberlain, C. P., J. R. Waldbauer, K. Fox-Dobbs, S. D. Newsome, P. L. Koch, D. R. Smith, M. E. Church, S. D. Chamberlain, K. J. Sorenson, and R. Riseborough. 2005. Pleistocene to recent dietary shifts in California condor. Proc. Nat. Acad. Sci. 102:16707–16711.
Chamberlain, D., M. Kibule, R. Q. Skeen, and C. Pomeroy. 2018. Urban bird trends in a rapidly growing tropical city. Ostrich 89:275–280.
Cody, M. L. 1974. Competition and the structure of bird communities. Princeton: Princeton University Press.
Cone, C. D., Jr. 1962. Thermal soaring of birds. Am. Sci. 50:180–209.
Cortés-Avizanda, A., R. Jovani, M. Carrete, and J. A. Donazar. 2012. Resource unpredictability promotes species diversity and coexistence in an avian scavenger guild: a field experiment. Ecology 93:2570–2579.
D'Elia, J., and S. M. Haig. 2013. California Condors in the Pacific Northwest. Corvallis: Oregon State University Press.
Dermody, B. J., C. J. Tanner, and A. L. Jackson. 2011. The evolutionary pathway to obligate scavenging in *Gyps* vultures. PLoS ONE 6:e24635.

DeVault, T. L., O. E. Rhode, Jr., and J. A. Shivik. 2003. Scavenging by vertebrates: behavioral, ecological and evolutionary perspective on an important energy transfer pathway in terrestrial ecosystems. Oikos 2003:225–234.

Duriez, O., A. Kato, C. Tromp, G. Dell'Omo, A. L. Vyssotski, F. Sarrazin, and Y. Ropert-Coudert. 2014. How cheap is soaring flight in raptors? A preliminary investigation in freely-flying vultures. PLoS ONE 9:e84887.

Dwyer, J. F., and S. G. Cockwell. 2011. Social hierarchy of scavenging raptors on the Falkland Islands, Malvinas. J. Raptor Res. 45:229–235.

Eitniear, J. C., and S. M. McGehee. 1994. Cannibalism by cathartid vultures. Vulture News 31:16–19.

Eitniear, J. C., and S. M. McGehee. 2017. Lesser Yellow-headed Vulture mandibular ecomorphology and feeding interactions at an established feeding site in Belize. Texas J. Sci. 69:39–48.

Fisher, H. J. 1944. The skulls of the Cathartid vultures. Condor 46:272–296.

Grilli, M. G., S. A. Lambertucci, J.-F. Therrien, and K. L. Bildstein. 2017. Wing size but not wing shape is related to migratory behavior in a soaring bird. J. Avian Biol. 48:669–678.

Hiraldo, F., and J. A. Donázar. 1990. Foraging time in the Cinereous Vulture *Aegypius monachus:* seasonal and local variations and influence of weather. Bird Study 37:128–132.

Houston, C. S., P. H. McLoughlin, J. T. Mandel, M. J. Bechard, M. J. Stoffel, D. R. Barber, and K. L. Bildstein. 2011. Breeding home ranges of migratory Turkey Vultures near their northern limit. Wilson J. Ornithol. 123:472–478.

Houston, D. C. 1974. The role of Griffon Vultures *Gyps africanus* and *Gyps rueppellii* as scavengers. J. Zool. Lond. 172:35–46.

Houston, D. C. 1976. Breeding of the White-backed and Rüppell's Griffon Vultures, *Gyps africanus* and *Gyps rueppellii.* Ibis 118:14–40.

Houston, D. C. 1984. Does the King Vulture *Sarcoramphus papa* use a sense of smell to locate food? Ibis 126:67–69.

Houston, D. C. 1985. Evolutionary ecology of Afrotropical and Neotropical vultures in forests. Ornithol. Monogr. 36:856–864.

Houston, D. C. 1986. Scavenging efficiency of Turkey Vultures in tropical forest. Condor 88:318–323.

Houston, D. C. 1988. Competition for food between Neotropical vultures in forest. Ibis 130:402–417.

Houston, D. C. 1993. The incidence of healed fractures to wing bones of White-backed and Rüppell's Griffon Vultures *Gyps africanus* and *G. ruepellii* and other birds. Ibis 135:468–470.

Humboldt, A. von, and A. Bonpland. 1820. Voyages aux régions équinoxiales du neouveau continent. Vol. 6. Paris: J. Smith.

Jackson, A. L., G. D. Ruxton, and D. C. Houston. 2008. The effect of social facilitation on foraging success in vultures: a modelling study. Biol. Lett. 4:311–313.

Kendall, C. J. 2014. The early bird gets the carcass: temporal segregation and its effects on foraging success in avian scavengers. Auk 131:12–19.

Kiff, L. F. 2000. The current status of North American vultures. *In* Raptors at risk, ed. R. D. Chancellor and B.-U. Meyburg. Berlin: World Work. Grp. Birds of Prey and Owls.

Kroever, A. L., and E. W. Grifford. Karok myths. Berkeley: University California Press.

Kruuk, H. 1967. Competition for food between vultures in East Africa. Ardea 55:171–193.

Kruuk, H. 1972. The spotted hyena. Chicago: University Chicago Press.

Lack, D. 1946. Competition for food by birds of prey. J. Anim. Ecol. 15:123–129.

Lack, D. 1971. Ecological isolation in birds. Cambridge: Harvard University Press.

Lack, D. 1976. Island biology. Berkeley: University California.

Lambertucci, S. A., K. L. Speziale, T. E. Rogers, and J. M. Morales. 2009. How do roads affect the habitat use of an assemblage of scavenging raptors? Biodivers. Conserv. 18:2063–2074.

Lambertucci, S. A., J. Navarro, J. A. Sanchez Zapata, K. A. Hobson, P. A. E. Alacón, G. Wiemeyer, G. Blanco, F. Hiraldo, and J. A. Donázar. 2018. Tracking data and retrospective analyses of diet reveal the consequences of loss of marine subsides for an obligate scavenger, the Andean Condor. Proc. R. Soc. Lond. B 285:20180550.

Lemon, W. C. 1991. Foraging behavior of a guild of Neotropical vultures. Wilson Bull. 103:698–702.

Lowery, G. H., and W. W. Dalquist. 1951. Birds from the state of Veracruz, Mexico. Lawrence: U. Kansas, Pub. Mus. Nat Hist. 3.

Mallon, J. M., K. L. Bildstein, and T. E. Katzner. 2016. In-flight turbulence benefits soaring birds. Auk 133:79–85.

Mandel, J. T., K. L. Bildstein, G. Bohrer, and D. W. Winkler. 2008. Movement ecology of migration in Turkey Vultures. Proc. Nat. Acad. Sci. 195:19102–19107.

Mare, N. V., W. D. Edge, R. G. Anthony, and R. Valburg. 1995. Sheep carcass availability and use by Bald Eagles. Wilson Bull. 107:251–257.

Margalida, A., J. Bertran, J. Boudet, and R. Heredia. 2004. Hatching asynchrony, sibling aggression and cannibalism in the Bearded Vulture *Gypaetus barbatus*. Ibis 146:386–393.

Martinez De Lecea, F., A. Hernando, A. Illana, and J. Echegaray. 2011. Cannibalism in Eurasian Griffon Vultures *Gyps fulvus*. Ardea 99:240–242.

McNab, B. K. 1988. Food habitats and the basal rate of metabolism in birds. Oecologia 77:343–349.

McNab, B. K. 2002. The physiological ecology of vertebrates. Ithaca: Cornell University Press.

Miller, A. H. 1949. Some ecologic and morphologic considerations in the evolution of higher taxonomic categories. *In* Ornithologie als biologische wissencraft, ed. E. Mayr. Hiedelberg: C. Winter.

Mundy, P., D. Butchart, J. Ledger, and S. Piper. 1992. The vultures of Africa. London: Academic Press.

Murphy, R. C. 1936. Oceanic birds of South America, vol I. New York: Macmillan.

Odum, E. P. 1959. Fundamentals of ecology. Second edition. Philadelphia: Saunders.

Ogutu, J. O., H.-P. Piepho, H. T. Dublin, N. Bhola, and R. S. Reid. 2008. Rainfall influences on ungulate population abundance in the Mara-Serengeti ecosystem. J. Anim. Ecol. 77:814–829.

Olea, P. P., and P. Mateo-Tomás. 2009. The role of traditional farming practices in ecosystem conservation: the case of transhumance and vultures. Biol. Conserv. 142:1844–1853.

Pennycuick, C. J. 1971. Gliding flight of the White-back Vultures *Gyps africanus*. J. Exp. Biol. 55:13–38.

Pennycuick, C. J. 1971. Control of gliding angle in Rüppell's Griffon Vulture *Gyps ruepellii*. J. Exp. Biol. 55:39–46.

Pennycuick, C. J. 1972. Soaring behavior and performance of some East African birds, observed from a motor-glider. Ibis 114:178–218.

Pennycuick, C. J. 1973. The soaring flight of vultures. Sci. Am. 229:102–109.

Pennycuick, C. J. 1983. Thermal soaring compared in three dissimilar bird species, *Fregata magnificens*, *Pelecanus occidentalis* and *Coragyps atratus*. J. Exp. Biol. 102:307–325.

Pennycuick, C. J. 1998. Field observations of thermals and thermal streets, and the theory of cross-country soaring flight. J. Avian Biol. 29:33–43.
Pennycuick, C. J. 2008. Modelling the flying bird. Amsterdam: Academic Press.
Petrides, G. A. 1959. Competition for food between five species of East African vultures. Auk 76:104–106.
Pomeroy, D. E. 1975. Birds as scavengers of refuse in Uganda. Ibis 117:69–81.
Portugal, S. J., C. Murn, and G. R. Martin. 2017. White-headed Vultures *Trigonoceps occipitalis* show visual field characteristics of hunting raptors. Ibis 159:463–466.
Rodríguez-Estrella, R., and L. Rodríguez-Estrella. 1992. Kleptoparasitism and other interactions of Crested Caracara in the Cape Region Baja California, Mexico. J. Field Ornithol. 63:177–180.
Ruxton, G. D., and D. C. Houston. 2003. Could *Tyrannosaurus rex* have been a scavenger rather than a predator? An energetic approach. Proc R. Soc. Lond. 270:731–733.
Schaller, G. B. 1972. The Serengeti lion. Chicago: University Chicago Press.
Scherub, S., G. Bohrer, M. Wikelski, and R. Weinzierl. 2016. Behavioral adaptations into thin air. Biol. Lett. 12:20160432.
Schlacher, T. A., S. Strydom, and R. M. Connolly. 2013. Multiple scavengers respond rapidly to pulsed carrion resources at the land-ocean interface. Acta Oecol. 48:7–12.
Schlee, M. A. 2007. King Vultures (*Sarcoramphus papa*) follow jaguar in the Serranía de la Cerbatana, Venezuela. Vulture News 57:4–16.
Snyder, N., and H. Snyder. 2000. The California Condor: a saga of natural history and conservation. San Diego: Academic Press.
Spiegel, O., W. M. Getz, and R. Nathan. 2013. Factors influencing foraging search efficiency: why do scarce Lappet-faced Vultures outperform ubiquitous White-backed Vultures. Am. Nat. 181:E102-E115.
Stewart, P. A. 1978. Behavioral interactions and niche separation in Black and Turkey Vultures. Living Bird 17:79–84.
Stolen, E. D. 1996. Black and Turkey Vulture interactions with Bald Eagle in Florida. Fl. Field Nat. 24:43–45.
Travaini, A. J., A. Donázar, A. Rodriguez, Olga Ceballos, M. Funes, M. Delibes, and F. Hiraldo. 1998. Use of European hare (*Lepus europaeus*) carcasses by an avian scavenging assemblage in Patagonia. J. Zool. 46:175–181.
Tristram, H. B. 1859. On the ornithology of northern Africa (Part 3). Ibis 1:415–435.
Virani, M. Z., A. Monadjem, S. Thomsett, and C. Kendall. 2012, Seasonal variation in Rüppell's Vultures *Gyps rueppellii* at Kwenia, southern Kenya and implication for conservation. Bird Conserv. Int. 22:260–269.
Wallace, M. P., and S. A. Temple. 1987. Competitive interactions within and between species in a guild of avian scavengers. Auk 104:290–295.

Chapter 5: Movement Behavior

Batbayar, N., R. Reading, D. Kenny, T. Natsagdorf, and P. Won Kee. 2008. Migration and movement patterns of Cinereous Vultures in Mongolia. Falco 32:5–7.
Bildstein, K. L., M. J. Bechard, P. Porras, E. Campo, and C. Farmer. 2007. Seasonal abundances and distributions of Black Vultures (*Coragyps atratus*) and Turkey Vultures (*Cathartes aura*) in Costa Rica and Panama: evidence for reciprocal migration in the Neotropics. *In* Hawkwatching in the America, ed. K. L. Bildstein, D. R. Barber, and A. Zimmerman. Kempton: Hawk Mountain Sanctuary.

Bildstein, K. L., M. J. Bechard, C. Farmer, and L. Newcomb. 2009. Narrow sea crossings present major obstacles to migrating *Gyps fulvus*. Ibis 151:382–391.

Buckley, N. J. 1999. Black Vulture (*Coragyps atratus*). Birds of North America, No. 411, ed. A. Poole and F. Gill. Philadelphia: Birds of North America.

Burt, W. H. 1943. Territoriality and home range concepts as applied to mammals. J. Mammal. 24:346–352.

Dabone, C., R. Buij, A. Oued, J. B. Adajakpa, W. Wendengoudi, and P. D. M. Weesie. 2019. Impact of human activities on the reproduction of Hooded Vultures *Necrosyrtes monachus* in Burkina Faso. Ostrich 90:53–61.

Duriez, O., R. Harel, and O. Hatzofe. 2019. Studying movement ecology of avian scavengers to understand carrion ecology. *In* Carrion ecology and management, ed. P. Olea, P. Mateo-Tomás, J. A. Sanchez Zapata, and José Antonio. Wildlife Research Monographs 2. Cham, Switzerland: Springer Nature.

Eisenmann, E. 1963. Is the Black Vulture migratory? Wilson Bull. 75:244–249.

García-Ripollés, C., P. López-López, and V. Urios. 2010. First description of migration and wintering of adult Egyptian Vultures *Neophron percnopterus* tracked by GPS satellite telemetry. Bird Study 57:261–265.

Gavashelishvili, A., M. McGrady, M. Ghasabian, and K. L. Bildstein. 2012. Movements and habitat use by immature Cinereous Vultures (*Aegypius monachus*) from the Caucasus. Bird Study 59:449–462.

Grinnell, J. 1922. The role of the "accidental." Condor 39:373–380.

Henckel, R. E. 1976. Turkey Vulture banding problem. N. Am. Bird Band. 1:126.

Hirschauer, M. T., K. Wolter, and W. Neser. 2016. Natal philopatry in young Cape Vultures *Gyps coprotheres*. Ostrich 2016:1–4.

Jallow, M., C. R. Barlow, L. Sanyang, L. Dibba, C. Kendall, M. Bechard, and K. L. Bildstein. 2016. High population of the critically endangered Hooded Vulture *Necrosyrtes monachus* in Western Region, The Gambia, confirmed by road surveys in 2013 and 2015. Malimbus 38:23–28.

Jenny, D., M. Kéry, P. Trotti, and E. Bassi. 2018. Philopatry in a reintroduced population of Bearded Vultures *Gypaetus barbatus* in the Alps. J. Ornithol. 159:507–515.

Jespersen, P., and A. V. Taning, ed. 1950. Studies in bird migration: being the collected papers H. Chr. C. Mortensen. Copenhagen: Munksgaard.

Kays, R., M. C. Crofoot, W. Jetz, and M. Wikelski. 2015. Terrestrial animal tracking as an eye on life and planet. Science 348:1222.

Kenny, D., N. Batayar, P. Tsolmonjav, M. Willis, J. Azua, and R. Reading. 2008. Dispersal of Eurasian Black Vulture *Aegypius monachus* fledglings from the Ikh Nart Nature Reserve, Mongolia. Vulture News 59:13–19.

Kirk, D. A., and A. G. Gosler. 1994. Body condition varies with migration and competition in migrant and resident South American vultures. Auk 111:933–944.

Kirk, D. A., and M. J. Mossman. 1998. Turkey Vulture (*Cathartes aura*). Birds of North America, No. 339, ed. A. Poole and F. Gill. Philadelphia: Birds of North America.

Le Gouar, P., A. Robert, J-P. Choisy, S. Henriquet, P. Lecuyer, C. Tessier, and F. Sarrazin. 2008. Roles of survival and dispersal in reproduction success of Griffon Vultures (*Gyps fulvus*). Ecol. Appl. 18:859–872.

López-López, P., C. García-Ripollés, and V. Urios. 2014. Individual repeatability and spatial flexibility of migration routes of trans-Saharan migratory raptors. Curr. Zool. 60:642–652.

López-López, P., I. Zuberogoitia, M. Alcántara, and J. A. Gil. 2013. Philopatry, natal dispersal, first settlement and age of first breeding of Bearded Vultures *Gypaetus barbatus* in central Pyrenees. Bird Study 60:555–560.

Meretsky, V. J., and N. F. R. Snyder. 1992. Range use and movements of California Condors. Condor 94:313–335.

Mundy, P., D. Butchart, J. Ledger, and S. Piper. 1992. The vultures of Africa. London: Academic Press.

Oppel, S., and 8 coauthors. 2015. High juvenile mortality in a declining population of a long-distance migratory raptor. Ibis 157:245–557.

Phipps, W. L., and 39 coauthors. 2019. Spatial and temporal variability in migration of a soaring raptor across three continents. Front. Ecol. Evol. 7:article 323.

Reading, R., G. Maude, P. Hancock, D. Kenny, and R. Garbett. 2014. Comparing different types of patagial tags for use on vultures. Vulture News 67:33–42.

Skutch, A. F. 1969. Notes on the possible migration and nesting of Black Vultures in Central America. Auk 86:726–731.

Snyder, N., and H. Snyder. 2000. The California Condor: a saga of natural history and conservation. San Diego: Academic Press.

Thompson, L. J., J. P. Davies, K. L. Bildstein, and C. T. Downs. 2017. Removal (and attempted removal) of nest material from a Hooded Vulture *Necrosyrtes monachus* nest by a starling and a Hooded Crow. Ostrich 88:183–187.

Thompson, L. J., J. P. Davies, M. Gudehus, A. J. Botha, K. L. Bildstein, C. Murn, and C. T. Downs. 2017. Visitors to nests of Hooded Vultures *Necrosyrtes monachus* in north-eastern South Africa. Ostrich 88:155–162.

Thompson, L. J., D. Barber, M. Bechard, A. J. Botha, K. Wolter, W. Neser, E. R. Buechley, R. Reading, R. Garbett, P. Hancock, G. Maude, M. Z. Virani, S. Thomsett, H. Lee, D. Ogada, C. Barlow, and K. L. Bildstein. 2020. Variation in monthly home ranges of Hooded Vultures *Necrosyrtes monachus* in western, eastern, and southern Africa. Ibis doi.org/10.1111/ibi.12836.

Wallace, M. P., P. G. Parker, and S. A. Temple. 1980. An evaluation of patagial markers for Cathartid vultures. J. Field Ornithol. 51:309–314.

Yosef, R., and D. Alon. 1997. Do immature Palearctic Egyptian Vultures *Neophron percnopterus* remain in Africa during the northern summer? Vogelwelt 118:285–289.

Chapter 6: Social Behavior

Batbayar, N., R. Reading, D. Kenny, T. Natsagdorf, and P. Won Kee. 2008. Migration and movement patterns of Cinereous Vultures in Mongolia. Falco 32:5–7.

Bildstein, K. L., M. J. Bechard, P. Porras, E. Campo, and C. Farmer. 2007. Seasonal abundances and distributions of Black Vultures (*Coragyps atratus*) and Turkey Vultures (*Cathartes aura*) in Costa Rica and Panama: evidence for reciprocal migration in the Neotropics. *In* Hawkwatching in the America, ed. K. L. Bildstein, D. R. Barber, and A. Zimmerman. Kempton: Hawk Mountain Sanctuary.

Bildstein, K. L., M. J. Bechard, C. Farmer, and L. Newcomb. 2009. Narrow sea crossings present major obstacles to migrating *Gyps fulvus*. Ibis 151:382–391.

Buckley, N. J. 1999. Black Vulture (*Coragyps atratus*). Birds of North America, No. 411, ed. A. Poole and F. Gill. Philadelphia: Birds of North America.

Burt, W. H. 1943. Territoriality and home range concepts as applied to mammals. J. Mammal. 24:346–352.

Dabone, C., R. Buij, A. Oued, J. B. Adajakpa, W. Wendengoudi, and P. D. M. Weesie. 2019. Impact of human activities on the reproduction of Hooded Vultures *Necrosyrtes monachus* in Burkina Faso. Ostrich 90:53–61.

Duriez, O., R. Harel, and O. Hatzofe. 2019. Studying movement ecology of avian scavengers to understand carrion ecology. *In* Carrion ecology and management, ed. P. Olea, P. Mateo-Tomás, J. A. Sanchez Zapata, and José Antonio. Wildlife Research Monographs 2. Cham, Switzerland: Springer Nature.

Eisenmann, E. 1963. Is the Black Vulture migratory? Wilson Bull. 75:244–249.

García-Ripollés, C., P. López-López, and V. Urios. 2010. First description of migration and wintering of adult Egyptian Vultures *Neophron percnopterus* tracked by GPS satellite telemetry. Bird Study 57:261–265.

Gavashelishvili, A., M. McGrady, M. Ghasabian, and K. L. Bildstein. 2012. Movements and habitat use by immature Cinereous Vultures (*Aegypius monachus*) from the Caucasus. Bird Study 59:449–462.

Grinnell, J. 1922. The role of the "accidental." Condor 39:373–380.

Henckel, R. E. 1976. Turkey Vulture banding problem. N. Am. Bird Band. 1:126.

Hirschauer, M. T., K. Wolter, and W. Neser. 2016. Natal philopatry in young Cape Vultures *Gyps coprotheres*. Ostrich 2016:1–4.

Jallow, M., C. R. Barlow, L. Sanyang, L. Dibba, C. Kendall, M. Bechard, and K. L. Bildstein. 2016. High population of the critically endangered Hooded Vulture *Necrosyrtes monachus* in Western Region, The Gambia, confirmed by road surveys in 2013 and 2015. Malimbus 38:23–28.

Jenny, D. M. Kéry, P. Trotti, and E. Bassi. 2018. Philopatry in a reintroduced population of Bearded Vultures *Gypaetus barbatus* in the Alps. J. Ornithol. 159:507–515.

Jespersen, P., and A. V. Taning, ed. 1950. Studies in bird migration: being the collected papers H. Chr. C. Mortensen. Copenhagen: Munksgaard.

Kays, R., M. C. Crofoot, W. Jetz, and M. Wikelski. 2015. Terrestrial animal tracking as an eye on life and planet. Science 348:1222.

Kenny, D., N. Batayar, P. Tsolmonjav, M. Willis, J. Azua, and R. Reading. 2008. Dispersal of Eurasian Black Vulture *Aegypius monachus* fledglings from the Ikh Nart Nature Reserve, Mongolia. Vulture News 59:13–19.

Kirk, D. A., and A. G. Gosler. 1994. Body condition varies with migration and competition in migrant and resident South American vultures. Auk 111:933–944.

Kirk, D. A., and M. J. Mossman. 1998. Turkey Vulture (*Cathartes aura*). Birds of North America, No. 339, ed. A. Poole and F. Gill. Philadelphia: Birds of North America.

Le Gouar, P., A. Robert, J-P. Choisy, S. Henriquet, P. Lecuyer, C. Tessier, and F. Sarrazin. 2008. Roles of survival and dispersal in reproduction success of Griffon Vultures (*Gyps fulvus*). Ecol. Appl. 18:859–872.

López-López, P., C. García-Ripollés, and V. Urios. 2014. Individual repeatability and spatial flexibility of migration routes of trans-Saharan migratory raptors. Curr. Zool. 60:642–652.

López-López, P., I. Zuberogoitia, M. Alcántara, and J. A. Gil. 2013. Philopatry, natal dispersal, first settlement and age of first breeding of Bearded Vultures *Gypaetus barbatus* in central Pyrenees. Bird Study 60:555–560.

McGahan, J. 1972. Behavior and ecology of the Andean Condor. Ph.D. diss., U. Wisconsin, Madison.

Meretsky, V. J., and N. F. R. Snyder. 1992. Range use and movements of California Condors. Condor 94:313–335.

Mundy, P., D. Butchart, J. Ledger, and S. Piper. 1992. The vultures of Africa. London: Academic Press.

Oppel, S., and 8 coauthors. 2015. High juvenile mortality in a declining population of a long-distance migratory raptor. Ibis 157:245–557.

Phipps, W. L., and 39 coauthors. 2019. Spatial and temporal variability in migration of a soaring raptor across three continents. Front. Ecol. Evol. 7:article 323.
Reading, R., G. Maude, P. Hancock, D. Kenny, and R. Garbett. 2014. Comparing different types of patagial tags for use on vultures. Vulture News 67:33–42.
Skutch, A. F. 1969. Notes on the possible migration and nesting of Black Vultures in Central America. Auk 86:726–731.
Snyder, N., and H. Snyder. 2000. The California Condor: a saga of natural history and conservation. San Diego: Academic Press.
Thompson, L. J., J. P. Davies, K. L. Bildstein, and C. T. Downs. 2017. Removal (and attempted removal) of nest material from a Hooded Vulture *Necrosyrtes monachus* nest by a starling and a Hooded Crow. Ostrich 88:183–187.
Thompson, L. J., J. P. Davies, M. Gudehus, A. J. Botha, K. L. Bildstein, C. Murn, and C. T. Downs. 2017. Visitors to nests of Hooded Vultures *Necrosyrtes monachus* in north-eastern South Africa. Ostrich 88:155–162.
Thompson, L. J., D. Barber, M. Bechard, A. J. Botha, K. Wolter, W. Neser, E. R. Buechley, R. Reading, R. Garbett, P. Hancock, G. Maude, M. Z. Virani, S. Thomsett, H. Lee, D. Ogada, C. Barlow, and K. L. Bildstein. 2020. Variation in monthly home ranges of Hooded Vultures *Necrosyrtes monachus* in western, eastern, and southern Africa. Ibis doi.org/10.1111/ibi.12836.
Wallace, M. P., P. G. Parker, and S. A. Temple. 1980. An evaluation of patagial markers for Cathartid vultures. J. Field Ornithol. 51:309–314.
Yosef, R., and D. Alon. 1997. Do immature Palearctic Egyptian Vultures *Neophron percnopterus* remain in Africa during the northern summer? Vogelwelt 118:285–289.

Chapter 7: Vultures and People

Alacón, P. A. E., and S. A. Lambertucci. 2018. Pesticides thwart condor conservation. Science 369:612.
Albert, L. A., C. Barcenas, M. Ramos, and E. Iñigo. 1989. Organochlorine pesticides and reduction of eggshell thickness in a Black Vulture *Coragyps atratus* population of the Tuxtla Valley, Chiapas, Mexico. *In* Raptors in the modern world, ed. B.-U. Meyburg and R. D. Chancellor. London: World Work. Grp. Birds of Prey and Owls.
Anderson, M. D., A. W. A. Maritz, and E. Oosthuysen. 1999. Raptors drowning in farm reservoirs in South Africa. Ostrich 70:139–144.
Angelov, I., I. Hashim, and S. Oppel. 2013. Persistent electrocution mortality of Egyptian Vultures *Neophron percnopterus* over 28 years in East Africa. Bird Conserv. Int. 23:1–6.
Auge, A. A. 2017. Anthropogenic debris in the diet of turkey vultures (*Cathartes aura*) in a remote and low-populated South Atlantic island. Polar Biol. 40:799–805.
Bickerton, D. 2009. Adam's tongue: how humans made language, how language made humans. New York: Hill and Wang.
Bijleveld, M. 1974. Birds of prey in Europe. London: Macmillan.
Bildstein, K. L., and J. F Therrien. 2018. Urban birds of prey: a lengthy history of human-raptor cohabitation. *In* Urban raptors, C. W. Boal and C. R. Dykstra, ed. Washington, D.C.: Island Press.
BirdLife International. 2020. Saving Africa's Vultures. https://www.birdlife.org/african-vultures (last accessed March 2020).

Breen, B. M., and K. L. Bildstein. 2007. Distribution and abundance of Turkey Vultures (*Cathartes aura*) in the Falkland Islands, summer 2006–2007 and autumn–Winter 2007. Unpubl. report. Stanley: Falklands Conservation.

Buij, R., G. Nikolaus, R. Whytock, D. J. Ingram, and D. Ogada. 2006. Trade of threatened vultures and other raptors for fetish and bushmeat in West and Central Africa. Oryx 50:606–616.

Burnett, L. J., and 9 coauthors. 2013. Eggshell thinning and depressed hatching success of California Condors reintroduced to central California. Condor 115:477–491.

Carneiro, M. A., P. A. Oliveira, R. Brandao, O. N. Francisco, R. Velarde, S. Levin, and B. Colaco. 2016. Lead poisoning due to lead pellet ingestion in Griffon Vultures (*Gyps fulvus*) from the Iberian Peninsula. J. Avian Med. Surg, 30:274–279.

Carpenter, J. W., O. H. Pattee, S. H. Fritts, B. A. Rattner, S. N. Wiemeyer, J. A. Royale, and M. R. Smith. 2003. Experimental lead poisoning in Turkey Vultures (*Cathartes aura*). J. Wildl. Dis. 39:96–104.

Couzens, D. 2010. Atlas of rare birds. Cambridge: MIT Press.

Cunningham, A. A., V. Prakash, D. Pain, G. R. Ghalsasi, G. A. H. Wells, G. N. Kolte, P. Nighot, M. S. Goudar, S. Kshirsagar, and A. Rahmani. 2003. Indian vultures: victims of an infectious disease epidemic? Anim. Conserv. 6:198–197.

Cushman, G. T. 2005. The most valuable birds in the world: international conservation science and the revival of Peru's guano industry, 1909–1965. Environ. Hist. 10:477–509.

Cuthbert, R., R. E. Green, S. Ranade, S. Saravanan, D. J. Pain, V. Prakash, and A. A. Cunningham. 2006. Rapid population declines of Egyptian Vulture (*Neophron percnopterus*) and Red-headed Vulture (*Sarcogyps clavus*) in India. Anim. Conserv. 9:249–254.

D'Elia, J., and S. M. Haig. 2013. California Condors in the Pacific Northwest. Corvallis: Oregon State University Press.

DeVault, T. L., O. E. Rhode, Jr., and J. A. Shivik. 2003. Scavenging by vertebrates: behavioral, ecological and evolutionary perspective on an important energy transfer pathway in terrestrial ecosystems. Oikos 102:225–234.

Galushin, V. M. 1971. A huge urban population of birds of prey in Delhi, India. Ibis 113:522.

Gangoso, L., P. Álvarez-Lioret, A. A. B. Rodríguez-Navarro, E. Mateo, F. Hiraldo, and J. A. Donázar. 2009. Long-term effects of lead poisoning on bone mineralization in vultures exposed to ammunition sources. Environ. Poll. 157:569–574.

Garbett, R., G. Maude, P. Hancock. D. Kenny, R. Reading, and A. Amar. 2018. Association between hunting and elevated blood lead levels in critically endangered African white-backed Vulture *Gyps africanus* Sci. Total Environ. 630:1654–1665.

Garcia-Fernandez, A. J., E. Martinez-Lopez, D. Romero, P. Maria-Mojica, A. Godino, and P. Jimenez. 2005. High levels of blood lead in Griffon Vultures (*Gyps fulvus*) from Cazorla natural park (southern Spain). Environ. Toxicol. 20:459–463.

Gilbert, M., R. T. Watson, S. Ahmed, M. Asim, and J. A. Johnson. 2007. Vulture restaurants and their role in reducing diclofenac poisoning in Asian vultures. Bird Conserv. Int. 17:63–77.

Grady, W. 1997. Vulture: nature's ghastly gourmet. San Francisco: Sierra Club.

Green, R. E., I. Newton, S. Schultz, A. A. Cunningham, M. Gilbert, D. J. Pain, and V. Prakash. 2004. Diclofenac poisoning as a cause of vulture population declines across the Indian subcontinent. J. Applied Ecol. 41:793–800.

Greenway, J. C., Jr. 1958. Extinct and vanishing birds of the world. Spec. Publ. 13. New York: Am. Comm. Int. Wild Life Protect.

Hernández, M., and A. Margalida. 2009. Poison-related mortality effects in an endangered Egyptian Vulture (*Neophron percnopterus*) population in Spain. Eur. J. Wildl. Res. 55:415–423.

Hirschfeld, E., A. Swash, and R. Still. 2013. The world's rarest birds. Princeton: Princeton University Press.

Houlihan, P. F. 1986. The birds of ancient Egypt. Warminster: Aris & Phillips.

Houston, D. C. 1974. The role of griffon vultures *Gyps africanus* and *Gyps rueppellii* as scavengers. J. Zool., London 172:35–46.

Houston, D. C. 1976. Breeding of the White-backed and Rüppell's Griffon Vultures, *Gyps africanus* and *Gyps rueppellii*. Ibis 118:14–40.

Houston, D. C. 1984. Does the King Vulture *Sarcoramphus papa* use a sense of smell to locate food? Ibis 126:67–69.

Houston, D. C. 1996. The effect of altered environments on vultures. *In* Raptors in human landscapes, ed. D. M. Bird, D. E. Varland, and J. J. Negro. London: Academic Press.

Houston, D. C., and J. E. Cooper. 1975. The digestive tract of the whiteback Griffon Vulture and its role in disease transmission among wild ungulates. J. Wildl. Dis. 11:306–313.

Houston, D. C., and S. E. Piper. 2006. Proceedings of the international conference on conservation and management of vulture populations, 14–16 November 2005, Thessaloniki, Greece. Heraklion, Crete: Natural History Museum of Crete and WWF Greece.

Houston, D. C., A. Mee, and M. McGrady. 2007. Why do condors and vultures eat junk? the implications for conservation. J. Raptor Res. 41:235–238.

Howes, C. G. 2016. Power line risk to Cape (*Gyps coprotheres*) and White-backed (*G. africanus*) Vultures in southern Africa. Johannesburg: Unpubl. Ph.D. diss., University of Witwatersrand.

Iñígo Elías, E. 1987. Feeding habits and ingestion of synthetic products in a Black Vulture population from Chiapas, Mexico. Acta Zool. Mexico 22:1–15.

Kalmbach, E. 1939. American vultures and the toxin of *Clostridium botulinum*. Am. Vet. Med. Assn. J. 54:187–191.

Kelly, T. R., P. H. Bloom, S. G. Torres, Y. Z. Hernandez, R. H. Poppenga, W. M. Boyce, and C. K. Johnson. 2011. Impact of California lead ammunition ban on reducing lead exposure in Golden Eagles and Turkey Vultures. PLoS ONE 6:e17656.

Kiff, L. F. 1989. DDE and the California Condor *Gymnogyps californianus*: the end of a story? *In* Raptors in the modern world, ed. B.-U. Meyburg and R. D. Chancellor. London: World Work. Grp. Birds of Prey and Owls.

Krüger, C. S., and A. Amar. 2018. Lead exposure in the critically endangered Bearded Vulture (*Gypaetus barbatus*) population in southern Africa. J. Raptor Res. 52:491–499.

Lack, D. 1976. Island biology. Berkeley: University of California Press.

Lambertucci, S. A., J. A. Donázar, A. D. Huertas, B. Jiménez, M. Sáez, J. A. Sanchez-Zapata, and F. Hiraldo. 2011. Biol. Conserv. 144: 1464–1471.

Linsdale, J. M. 1932. Further facts concerning losses to wild animal life through pest control in California. Condor 34:121–135.

Loveridge, R., and 11 coauthors. 2019. Poisoning causing the decline in South-East Asia's largest vulture population. Bird Conserve. Int. 29:41–54.

Lyell, C. 1830–1833. Principles of geology, 3 vols. London: John Murray.

Mander, M., N. Diedruchs, L. Ntuli, K. Mavundla, V. Williams, and S. McKean. 2007. Survey of the trade in vultures for the traditional health industry in South Africa. Pietermaritzburg: Unpubl. Report to Ezemvelo KNZ Wildlife.

Margalida, A., R. Heredia, M. Razin, and M. Hernández. 2008. Sources of variation in mortality of the Bearded Vulture *Gypaetus barbatus* in Europe. Bird Conserv. Int. 18:1–10.

Martin, G. R., S. J. Portugal, and C. P. Murn. 2012. Visual fields, foraging and collision vulnerability in *Gyps* vultures. Ibis 154:626–631.

Mateo-Tomás, P., P. Olea, and J. V. López-Bao. 2018. Europe's uneven laws threaten scavengers. Science 360:613–614.

May, R. B. 1935. The hawks of North America. New York: National Assoc. Audubon Soc.

McKean, S., M. Mander, N. Diedrichs, L. Ntuli, K. Mavundla, V. Williams, and J. Wakelin. 2013. The impact of traditional use on vultures in South Africa. Vulture News 65:15–36.

Mee, A., and L. S. Hill. 2007. California condors in the 21st century. Studies in ornithology, No. 2. Washington D.C.: Nuttal Ornithol. Cl. and Am. Ornithol. Union.

Mee, A., B. A. Rideout, J. A. Hamber, J. N. Todd., G. Austin, M. Clark, and M. P. Wallace. 2007. Junk ingestion and nestling mortality in an introduced population of California Condors *Gymnogyps californianus*. Bird Conserv. Int. 17:119–130.

Mingozzi, T., and R. Estève. 1997. Analysis of a historical extirpation of the Bearded Vulture (*Gypaetus barbatus*) (L.) in the western Alps (France-Italy): former distribution and causes of extirpation. Biol. Conserv. 79:155–171.

Mountfort, G. 1988. Rare birds of the world: a Collins/ICBP handbook. London: Penguin.

Mundy, P., D. Butchart, J. Ledger, and S. Piper. 1992. The vultures of Africa. London: Academic Press.

Mundy, P. J., K. I. Grant, J. Tannock, and C. L. Wessels. 1982. Pesticide residues and eggshell thickness of Griffon Vulture eggs in southern Africa. J. Wildl. Manage. 46:769–773.

Murphy, R. C. 1925. Bird islands of Peru. New York: Putnam's Sons.

Murphy, R. C. 1936. Oceanic birds of South America, Vol. I. New York: Macmillan Press.

Murphy, R. C. 1936. Oceanic birds of South America, Vol. 2. New York: Macmillan Press.

Oaks, J. L., B. A. Rideout, M. Gilbert, R. Watson, M. Virani, and A. A. Khan. 2001. Summary of diagnostic investigation into vulture mortality: Punjab Province, Pakistan, 2000–2001. *In* Reports from the workshop on Indian *Gyps* vultures, ed. T. Katzner and J. Parry-Jones. Gloucestershire: Natl. Centre Birds of Prey.

Oaks, J. L., M. Gilbert, M. Z. Virani, R. T. Watson, C. U. Meteyer, B. A. Rideout, H. L. Shivaprasad, S. Ahmed, M. J. I. Chaudhry, M. Arshad, S. Mahmood, A. Ali, and A. A. Khan. 2004. Diclofenac residues as the cause of vulture population decline in Pakistan. Nature 427:630–633.

Ogada, D. L. 2014. The power of poison: pesticide poisoning of Africa's wildlife. Ann. New York Acad. Sci. 1322:1–20.

Ogada, D., A. Botha, and P. Shaw. 2016. Ivory poachers and poisoning: drivers of Africa's declining vulture populations. Oryx 50:593–596.

Olea, P. P., and P. Mateo-Tomás. 2009. The role of traditional farming practices in ecosystem conservation: the case of transhumance and vultures. Biol. Conserv. 142:1844–1853.

Pain, D., A. A. Cunningham, P. F. Donald, J. W. Duckworth, D. C. Houston, T. Katzner, J. Parry-Jones, C. Poole, V. Prakash, P. Round, and R. Timms. 2003. Causes and effects of temporospatial declines of *Gyps* vultures in Asia. Conserv. Biol. 17:661–671.

Parmalee, P. W. 1954. The vultures: their movements, economic status and control in Texas. Auk 71:443–453.

Pattee, O. H., J. W. Carpenter, S. H. Fritts, B. A. Rattner, S. N. Wiemeyer, J. A. Royale, and M. R. Smith. 2006. Lead poisoning in captive Andean Condors (*Vultur gryphus*). J. Wildl. Dis. 42:772–779.

Piper, S. E. 2004. Vulture restaurants. *In* The vultures of southern Africa—Quo Vadis? ed. A. Monadjem, M. D. Anderson, S. E. Piper, and A. F. Boshoff. Johannesburg: Endangered Wildl. Trust Birds Prey Work. Grp.

Prakash, V. 1989. The general ecology of raptors in Keoladeo national park, Bharatpur. Bombay: Unpubl. Ph.D. diss.

Prakash, V. 1999. Status of vultures in Keoladeo National Park, Bharatpur, Rajasthan, with special reference to population crash in *Gyps* species. J. Bombay Nat. Hist. Soc. 96:365–378.

Prakash, V., and 8 coauthors. 2017. Recent changes of populations of critically endangered *Gyps* vultures in India. Bird Conserv. Int. 29:55–70.

Rao, K. M. 1992. Vultures endangered in Guntur and Prakasam districts (AP) and vulture eating community. Newsletter Birdwatchers 32:6–7.

Satheesan, S. M., and M. Satheesan. 2000. Serious vulture-hits to aircraft over the world. Amsterdam: Internal Bird Strike Committee Meeting.

Simons, D. D. 1986. Interactions between California Condors and humans in prehistoric far western North America. *In* Vulture biology and management, ed. S. R. Wilbur and J. A. Jackson. Berkeley: University of California Press.

Simmons, R. E., and A. R. Jenkins. 2007. Is climate change influencing the decline of Cape and Bearded Vultures in southern Africa? Vulture News 56:41–51.

Snyder, N. F. R., and H. A. Snyder. 2000. The California Condor: a saga of natural history and conservation. London: Academic Press.

Torres-Mura, J. C., M. L. Lemus, and F. Hertel. 2015. Plastic material in the diet of the Turkey Vulture (*Cathartes aura*) in the Atacama Desert, Chile. Wilson J. Ornithol. 127:134–138.

van Dooren, T. 2011. Vulture. London: Reaktion.

Van Wyk, E., H. Bouwman, H. van der Bank, G. H. Verdoorn, and D. Hofmann. 2001. Persistent organochlorine pesticides detected in blood and tissue samples of vultures from different localities in South Africa. Comp. Biochem. Physiol. Part C 129:243–264.

Virani, M. Z., C. Kendall, P. Njoroge, and S. Thomsett. 2010. Major declines in the abundance of vultures and other scavenging raptors in and around the Masai Mara ecosystem, Kenya. Biol. Conserv. 44:746–752.

Wiemeyer, G. M., M. Aa Perez, L. T. Bianchini, L. Sampietro, G. F. Brvo, N. L. Jacome, V. Astore, and S. A. Lambertucci. 2016. Repeated conservation threats in the America: high levels of blood and bone lead in Andean Condors widen the problem to a continental scale. Environ. Pollut. dx.doi.org/10.1016/j.env.pol.2016.10.025.

Zimmerman, G. S., B. A. Millsap, M. L. Avery, J. R. Sauer, M. C. Runge, and K. D. Richkus. 2019. Allowable take of Black Vultures in the eastern United States. J. Wildl. Manage. 83:272–282.

INDEX